Radical Principles

RADICAL PRINCIPLES

reflections of an

unreconstructed democrat

MICHAEL WALZER

Basic Books, Inc., Publishers

NEW YORK

The author is grateful for permission to reprint material from the following sources:

From *The Rise and Fall of the City of Mahagonny*, by Bertolt Brecht, translated by W. H. Auden and Chester Kallman. Original copyright © 1955 by Suhrkamp Verlag, Berlin. Translation copyright © 1960, 1976 by the Estate of W. H. Auden and Chester Kallman. Reprinted by permission of David R. Godine, Publishers, Inc.

From *Brecht: The Man and His Work*, by Martin Esslin. Copyright © 1959, 1960 by Martin Esslin. Reprinted by permission of Doubleday & Company, Inc.

From "Howl," *Howl and Other Poems*, by Allen Ginsberg. Copyright © 1956, 1959 by Allen Ginsberg. Reprinted by permission of City Lights Books.

From "The Poor," *Collected Earlier Poems*, by William Carlos Williams. Copyright 1938 by New Directions Publishing Corporation. Reprinted by permission of New Directions Publishing Corporation.

Library of Congress Cataloging in Publication Data

Walzer, Michael.
   Radical principles.

   Includes index.
   1. Democracy—Addresses, essays, lectures. 2. Liberalism—Addresses, essays, lectures. 3. Socialism—Addresses, essays, lectures. I. Title.
JC423.W292   320.5   79–56371
ISBN: 0–465–06824–3

FOR

Irving Howe

Lewis Coser

Stanley Plastrik

Manny Geltman

and all my comrades on *Dissent*

CONTENTS

Contents

Radical Principles

Introduction:
Radical Principles

Written over a period of fifteen years, these essays reflect, I
think, a more or less coherent political perspective. Still, they
are separate essays, stimulated by particular events, written
for particular occasions, and whatever coherence they have
does not take the form of a consecutive argument. Nor do
they reflect some deep theory of historical development or
social structure. I have ideas about both, but I don't have a
theory. On the Left, one is accustomed to apologize for defi-
ciencies of this sort because world-historical theory is generally
taken to be the essential prerequisite of political commentary.
Social life is one long series of interconnections, from the
division of labor in ancient Babylonia to the latest strike in
Bolivia, and unless one understands it all, one understands
nothing at all. I don't believe that, though I take theory seriously and have spent many years studying and teaching it.
The give and take of democratic politics invites another sort
of argument, more immediate, short-term, tentative. I am not
talking about campaign oratory and position papers—the
byproducts, sometimes the debris, of political argument—but

about those reflections for which the essay is the most appropriate form: focused on decades rather than centuries, on individual and group experience rather than historical process, on values rather than integral ideologies. These, then, are political essays. They are concerned with the immediate aspirations and the permanent principles that shape my own politics. Even the "Theory of Revolution" is a political essay, the theoretical analysis little more than a prolegomena to a defense of democracy.

Political argument has always co-existed with deep theory; the distance between them, the degree of linkage varies over time. Today, however, the relation of the two is peculiarly uncertain. None of the standard theories accounts for our everyday experience. This is true even of the one that has always seemed to me the most persuasive, that is, the Marxist theory of class conflict. I don't doubt that our collective future can still be projected in terms of the rise and fall of social classes. But the projections are murky, and the discoveries of "new" classes in contemporary sociological writing are mostly, as it were, discoveries made to order. No theorist has yet returned from his explorations, like a modern Columbus, with news of a new continent, unexpected but most certainly there. I also don't doubt that class analysis has a great deal to tell us about the actual constraints that shape our working lives, our family relationships, our politics and culture. But though there is significant light to be shed, it is hard to believe in any more complete illumination from Marxist theory, either in its original form or in any of its later political or academic revisions. Contemporary experience belies all such hopes. The sheer flux of events, the outbursts of irrational savagery, the long wars, the failure of working class parties to produce socialist societies, the depth and intensity of nationalist feeling, the drift toward authoritarianism: none of this can be accounted for in terms of the rhythms of rise and fall or in terms of a single set, even a complex set, of class identities and determinations.

Nor do any of the available Marxisms tell us enough about our own world, the world of advanced liberalism and advanced capitalism. Its values are largely unmeasured, its costs unanticipated. Bourgeois political culture has proven to be far more attractive and far more open than Marxists thought it would be. They consistently underestimated the significance of its two most important achievements: legal equality and legitimate opposition. Because of these two, workers were able to fight their way into the bourgeois state, which they transformed into a welfare state. And because of these two, the welfare state was able to incorporate workers into its institutional structures. All this was unexpected, and so were the individualizing effects of the incorporation, the long process by which merchant citizens and worker citizens were transformed into private consumers. And again, the seemingly endless provision of things to be consumed—material goods, welfare services, cultural artifacts of every conceivable sort—this too was unexpected.

And finally, liberation was unexpected—I mean, the breakdown of all those constraints, as old as the constraints of class, imposed on individuals because of their religion, race, or sex. For years, we (on the Left) were taught that the repression of religious and racial minorities, and the repression of women too, was inherent in the capitalist system, crucial to the maintenance of the reserve army and the reproduction of the work force. Repression would end only with the revolution, for the bourgeoisie, despite its general commitment to liberalism, was specifically committed to this or that form of bigotry, racism, and patriarchy. So it was, and so many of its members still are; and so was and is the working class. Racial, religious, and sexual chauvinism: none of these were ever class-bound feelings. But today, all such feelings, and the laws and customs that go with them, are at odds with the drift of our liberal culture, and we can begin to see that the long-term force of liberation will overcome them, as it will overcome every constraint on the right of individuals "to live their own

5

lives" without regard to color or sex or sexual preference or religious or political belief.

I don't want to overstate the case. The basic rights that are here at issue remain to be vindicated or fully vindicated. In the United States right now, women don't yet enjoy the opportunities that men take for granted; blacks still suffer from a heritage of racism. There are important and difficult battles still to be fought. And it remains conceivable, though I think it increasingly doubtful, that a full liberation—equality under the law, career open to talents, moral and cultural *laissez faire*—will require radical changes in the structure of liberalism and capitalism. On the other hand, at the further reaches of liberation, we already have extraordinary effects, without radical changes. And it's worth dwelling on these for a moment to see if we can catch a glimpse of our liberated future.

Consider, for example, our recently won right to watch "live sex acts" on stage: this is, I think, freedom of speech (since the sex is or might be symbolic); more generally, it is liberty and *laissez faire* in sexual life. Or consider the recent decision of a federal judge that a religious sect that "worships" the illegal drug peyote may distribute the drug to its members (it remains illegal for everyone else): this is, obviously, religious freedom. Or consider several recent judicial decisions permitting children to divorce their parents and parents to divorce their children: this is equal protection of the laws. All these are bizarre, though not illogical, examples of liberation. They neither contradict nor threaten the basic structure of liberal doctrine: here is individualism with a vengeance. But they inspire a bleak vision. I imagine a human being thoroughly divorced, freed of parents, spouse, and children, watching pornographic performances in some dark theater, joining (it may be his only membership) this or that odd cult, which he will probably leave in a month or two for another still odder. Is this a liberated human being? Well, one might say, he is an extreme case. But the tendencies that I have brought to a point in him are visible more generally. These tendencies are hardly likely to produce the autonomous men and women of bourgeois moral philosophy, and no more likely to produce the resolute revolutionary of Marxist theory.

The liberated human being looks rather like those unhappy members of "mass society" described in the social thought of the 1950s: dissociated, passive, lonely, ultimately featureless. That portrait was probably too starkly drawn; it missed, for one thing, the sheer recalcitrance of class. Liberation as we know it today co-exists with material inequalities of a highly structured sort. But it has effects independent of this structure. The most important effect is an odd doubling of conformist and eccentric behavior. The eccentricity is probably more indicative of contemporary mores than theorists of the fifties acknowledged. It represents a kind of escape from conformity for especially bold or especially disturbed individuals —and a diversion for the rest of us. At the same time as the general norms of social conduct are tightened, the margins of society are widened. We are, in fact, nowhere nearly so permissive as conservative writers sometimes suggest, but we are wonderfully tolerant of idiosyncratic tastes and strange beliefs, of cultic practices and deviant behavior—so long as we see nothing more in any of these than men and women in search of (private) satisfaction. We are even respectful of conscience, especially of religious conscience, if sometimes in a rather trivializing way. So that I am free, for example, to celebrate a black mass in my living room. And if I am unlikely to escape television coverage, my neighbors will complain only about the cameramen on their lawns; they won't say a word about my "religious preferences."

But it is perhaps more important to stress the new patterns of conformity, which are increasingly shaped to a set of rules different from any we have known before. At the very center of our social life, where satisfaction is sought in conventional ways, a certain sort of classless promiscuity has become one of the conventions. It is as if we were suddenly transported to Bertolt Brecht's city of Mahagonny, where:

> One means to eat all you are able;
> Two to change your loves about;
> Three means the ring and gaming table;
> Four, to drink until you pass out.
> Moreover, better get it clear
> That Don'ts are not permitted here.
> Moreover, better get it clear
> That Don'ts are not permitted here![1]

Brecht's list may seem a bit old-fashioned, even unimaginative; we could easily add to the list of abolished prohibitions. But the conception of human existence is instantly recognizable: a frenetic rush from one activity to another, one relationship to another, one sensation to another—a grim parody of Jefferson's pursuit of happiness.

The promiscuous use of people and things, the philosophy of "Do it!" (the phrase is Brecht's)—these are often taken to be the moral reflections of affluence and mass production, as if washing machines, vacuum cleaners, and ready-made clothing corrupted the human spirit, or as if the spirit were already corrupt and now suddenly released from the uplifting discipline of physical hardship. That's not Brecht's view. He thought Mahagonny a prototypical bourgeois city, stripped of all ideological veils, its moral character a reflection of the capitalist ethos. But that's not right either. Mahagonny is a liberated bourgeois city—not a sweatshop, or a marketplace, but an amusement park. The liberation is no less real for the fact that some people are still making money out of other people. And no less real for the fact that the pleasures offered are ultimately unsatisfying (they are not unsatisfying all along the way . . .). What is crucial about the gaming rooms of the city is the social range of their visitors. So long as no one is excluded, so long as everyone has a right to join in, and actually joins in, the realities of the experience are

[1] *The Rise and Fall of the City of Mahagonny*, trans. W. H. Auden and Chester Kallman (Boston, 1976), p. 68.

not adequately described with the adjectives of class. One has a sense of a city straining beyond class. Hence the feeling of liberation. At the same time, however, the effort is distorted, and the play corrupted, by the ancient habits of deference, passivity, and subordination. These are never overcome in the city of Mahagonny because they have their origins and their permanent base not in the sphere of leisure and consumption but in the sphere of work and politics. In Mahagonny, as in our own Las Vegas, the factory and the forum are missing.

Similarly, the welfare state, though it represents (as I shall argue again and again in these essays) an enormous political achievement and generates its own workaday politics, does not by itself produce either a community of workers or a community of citizens. It carries us beyond the classic structures of bourgeois society but not yet into a socialist society. So we are stranded, as it were, between two theoretical visions. The terrain of our lives is unmapped, our movement across it uncertain. Historically, this is a moment of opportunity, seized by all those individuals and groups (most of them undreamt of, or at least unmentioned, in Marxist theory) who assert their rights, discover their roots, come out of their closets, and do their things. I wish them well, and feel no sympathy with one of the more common responses to their appearance—the covert, inviting glance over one's shoulder at the policemen waiting in the wings. But liberation, if it requires just this moment of instability, must at some point take on stable forms, must give rise to coherent groups, lasting institutions, patterns of cooperative activity. What will these be like? I don't think anyone knows.

In such a time, in the city of Mahagonny, it is more important to have principles than to have theories. Now, one way to have principles is simply to be old-fashioned, curmudgeonly, reactionary: to endorse and defend the world of pre-liberation. One can yearn for the old stability, hierarchy, deference, economic discipline, and for the toughened, stoical men and women these (sometimes) produced; and one can attempt a principled resistance to each new wave of liberation. Such a response lacks generosity, though not necessarily courage, but it has, I think, a fundamental defect: its underlying principles can hardly be defended effectively without calling in the policemen from the wings. One might say of the Right today what Burke said of the revolutionary Left of his own time: "In the groves of *their* academy, at the end of every vista, you see nothing but the gallows."[2]

And yet, the deep principles of the Left also have their origins in the pre-liberated world. Where else could they have their origins? The long political and economic struggles could never have been sustained without common convictions, without an intimately shared moral sense. And we did not invent the convictions we shared. They were and are part of our cultural heritage: *our* heritage especially, because the old principles stood so often in sharp tension with the old political and economic order. The list is well-known: individual freedom, dignity, responsibility, equality, mutual respect, hard work, craftsmanship, honesty, and loyalty. And two more, less commonly acknowledged, authority and property. About these latter two I need to say a word, for in the city of Mahagonny they are both casually denied and secretly submitted to—a bad combination. The goal of democrats and socialists is to share and legitimize, but not to abolish authority. In the political movements of the Left, and in the future society too, it is crucial that some men and women be able to exercise authority and that others, despite their new and often touchy dignity, be willing to accept it. Even if offices were rotated among the membership or the citizenry, they would still have to be filled at any given moment by particular persons, who must be ready to do the work that needs to be done and equally ready to accept responsibility if they fail.

[2] *Reflections on the Revolution in France*, ed. Conor Cruise O'Brien (Harmondsworth, England, 1968), pp. 171–72.

Again, the goal of democrats and socialists is to share and legitimize, but not to abolish property. This too is best understood concretely, out of our common experience. No movement of the Left can possibly succeed unless it is viewed by its members as a *res publica,* a common possession, for whose treasury, offices, equipment, publications, and so on, all of them are responsible. Where private ownership is inappropriate or unjust, communal ownership takes its place: the things we make and use cannot be unowned. It hardly represents a cultural advance to undervalue these things, as if they were easily had, easily lost or discarded, and easily replaced. Under conditions of high affluence, individuals can perhaps sustain such a sensibility, but it is disastrous in parties, clubs, factories, and communes. We cannot value work without valuing the values that work produces.

These principles are enormously important. Every activist knows in his gut how critical they are, even when his head, wracked by ideologies, refuses agreement. That's why the counterculture of the late sixties, among leftists, was so much a matter of the head: it denied the deepest intimations of our political experience. Without these principles, the movement will be slack and listless or, alternatively, it will be frantic and hysterical, and in neither case can one imagine authority and property being shared for long. A slack socialism, without responsible, hard-working, and loyal citizens, would soon be converted into a new tyranny.

No one committed to principles of this sort is likely to feel comfortable in the city of Mahagonny. We have another sort of city in mind. It is, of course, a city in which liberated men and women might find a home. But we don't say of our homes that "Don'ts are not permitted here." Constraint is a necessary feature of any social world in which we also feel a sense of belonging, mutual respect, and shared responsibility. What I have called the bizarre forms of liberation represent a kind of liberalism *in extremis.* They are perhaps the fringe benefits of liberation struggles, but they also undermine the coherence

and solidarity that make struggle possible. Of many particular examples of liberation, it can properly be said that they
may or may not be realized in a new society: that will depend
upon the character of the society and its common life. Political freedom, indeed, is an absolute value, for without that
men and women with different ideas and interests cannot
share a common life. Religious toleration and cultural diversity are necessary corollaries of mutual respect. But "Do it!"
is not our morality, and we are prepared to tamper with the
creed of *laissez faire*. Individual liberty is meaningless until
it is incorporated within particular forms of social life, meaningless until it takes on shape and limit.

This incorporation is the socialist project. Socialism is the
effort to sustain older values within a social structure that
accommodates liberated, that is, free and equal individuals.
Nowhere is the importance of the effort more apparent than
in the city of Mahagonny. The working and living conditions
of early capitalism drove men and women, in desperation and
anger, to demand the immediate assistance of state officials.
And wherever those or parallel conditions reappear, the same
demand will be heard. But life in the contemporary amusement park generates different imperatives. Here it's not state
action that one longs for, but individual and group initiative.
And the politics that is necessary is not only the politics of
welfare, but also the *politics of politics*. Liberation will be an
empty experience, and a brief one, if it doesn't have as one
of its consequences the seizure of power and the acceptance
of responsibility by liberated men and women. They themselves must determine the shape of their common life: of the
work they do together and not only of the pleasures they privately pursue.

Hence socialists are advocates of community. I would stress
that this is not, most importantly, for the sake of intimacy and
good fellowship. From the ancient Greeks, we have learned
that politics is the enterprise of friends. But in any strong
sense of that word, I doubt that the citizens of the *polis* were

friends, each one of them to all the others. Friendship, like love, describes a more personal relation, and it is probably a mistake to seek the special delights of that relation in the public arena. Certainly, we can have associates, colleagues, co-workers, fellow members, even comrades, with whom we are not particularly friendly. No, when we talk about community, and then about decentralization, localism, workers' control, and so on, it is not because we are eager for warmth. Politics, after all, is an experience of conflict and hostility as well as an experience of cooperation. We seek community for the sake of knowledge and self-management. A secular view dominates our thought. Communion is for ritual occasions; what we want on an ordinary basis is space and shelter to put our principles into play. In our minds, freedom and dignity are not only norms for the treatment of individuals; they are also ideals which individuals themselves must embody and express. Socialists have an athletic conception of value. In the human fiber, morality is like muscle: it withers with disuse. And then communities are simply arenas for economic and political activity.

We seek communities, then, of a certain sort, not of any sort. Our goal is not an ecstatic union of the faithful, or a band of brethren bound to some charismatic leader, or a hierarchy of benevolent masters and docile servants. Warmth can be had in all of these, but in none of them are the arrangements of the common life open to popular scrutiny and revision. In none of them do individual members share political responsibility. In none of them, indeed, is there room for the claims and counter-claims of members who have learned to think of themselves as individuals. Only a democratic and egalitarian community can accommodate liberated men and women.

But democracy and equality are not the automatic products of liberation, at least not of liberation as we know it here in the intermediate zone, the unmapped terrain, between the bourgeois and socialist visions. Individuals set free to pursue happiness in their cultural and sexual lives—and in their leisure time—are not thereby brought together for cooperative activity. On the contrary: their self-determinations are private, and in that part of their lives that is socially structured and controlled, they are as likely as they ever were to accept the determinations of others. They are liberated on the side, as it were, still subordinate in the political and economic center of their lives.

That subordination has an old and familiar pattern, shaped above all by the lines of class. In contemporary America, discussions about social structure and stratification focus mostly on the question of mobility. Class, in the conventional view, is a prison, and our only concern is to facilitate individual escapes: another example of the politics of liberation. The public school was once thought to be the chief means of escape; today students of the educational process are less optimistic. But I wonder if they have not gotten the project wrong. For class is not only a prison; it is also a way of life, and surely the most important goal of educational policy, and of social policy more generally, should be to strengthen that common life, to open paths for class initiative and collective advance, not, or not only, to select children for individual advance. From a political standpoint, the most depressing contemporary statistics are not those that measure mobility (and show rather more of it than in many other societies), but those that measure organizational membership and participation. Here the correlations are stark. Men and women of the subordinate classes (however these are defined: the usual indices refer to income and education) are far less active in clubs, unions, churches, and political parties than are their better-off counterparts. Perhaps they are more tied to family and neighborhood; those ties should not be undervalued. But what is decisive politically is that they have less experience of voluntary association, committee work, public debate, and decision making. They are habituated to acquiesence, ready to accept the decisions of others, too often frightened, withdrawn, and passive. And I don't think there is much evidence that liberation, though it makes for new forms of ethnic and sexual assertiveness, has significantly altered the deeper habituations of class.

The result is the characteristic politics of the intermediate zone, marked by heightened demands for state action on behalf of minority groups and liberated individuals but not by any more fundamental social initiatives or institutional transformations. In these circumstances, the call for community is rarely concrete and pointed, rarely as precise as political argument ought to be. One too often hears community defended in the tone of voice one associates with the wringing of hands. Perhaps that is the tone of some of these essays too, though I have tried to avoid all the vague references to an idyllic past, now lost forever, that are so common in contemporary political literature. It is true that rates of political participation in the United States were once much higher than they are today, and that local communities were once much stronger. It is also true that the citizens of the Greek *polis* and the medieval commune often had a public life more vital than that of ordinary Americans. But our society is more complex, more heterogeneous, and above all more inclusive than nineteenth-century America, or any earlier political order, ever was. The mechanisms of integration and coordination are inevitably complex, hard to grasp, harder to change. In the United States today, community will have characteristically modern forms, or it won't exist at all.

Perhaps it won't exist at all. A new articulation of groups, a new division of powers: this may not be possible in the face of the modern omni-competent state. It is, to be sure, one of the clichés of public discourse these days that the state can't do everything and that there are evils beyond political reach. Like most clichés, the statements are true, and yet the range of state action continues to expand. A great deal of the political activity in which men and women on the Left are currently engaged, and rightly engaged, would require further expansion. I don't want to deny the necessary work of the state; I don't know how to do that. Contemporary denials, Right and Left, sound too much like a political re-enactment of Ludd-ism. They are anguished enough, but not serious enough. We can hardly avoid joining the debates about what state policy should be like, first in this area and then in that one. What patterns of distribution, what kinds of regulation and control, should state officials seek? Indeed, those of us who have been trained in the movements of the old Left or in the modern academy (the two sorts of training are not entirely dissimilar) are likely to think that there are right answers to such ques-tions—a single set of right answers, which state officials must in turn be taught. I am inclined to be skeptical about that. At least, in the intermediate zone that we now inhabit, the right answers are painfully unclear.

We need to think more about the procedures by which questions are answered, less about the answers themselves. Now, a concern with "procedural justice" is commonly taken to be the hallmark of liberalism; socialists, we are often told, are concerned with substance, with actual outcomes and con-crete effects. But this distinction is wrongheaded. Arguments about procedure are also arguments about the distribution of decision-making power, and this is a substantive matter. Who should make political decisions? At what level of political organization? With what sorts of committee work, consulta-tion, and voting? How should interest groups and parties be involved? What sorts of appeals should be permitted against decisions thought to be unjust? These are the most important questions. These are the questions that challenge established powers and well-worn patterns of control and obedience. And to these questions it seems clear that there is no single set of answers. One would hope for a wide range of experimenta-tion. But socialists, at least, will have an end in view against which to judge the success of the experiments. We will aim at a greater and greater degree of shared responsibility for common enterprises, at more participation, more initiative, a more lively sense of ownership and efficacy, extending across existing class lines.

And then, let policy outcomes be what they will be. Outcomes are crucial only when men and women are, so to speak, waiting for them, standing by as spectators, recipients, and consumers, or absorbed in their private lives, watching warily for threats and opportunities. Of course, we are all spectators some of the time and private individuals much of the time, and so political decisions will always have the aspect of things that happen to us. But insofar as we are involved in decision making, each particular outcome will also look like part of a process over which we can still exercise some general control. Democracies are self-sustaining and self-renewing. The continual debate, the ever-present opposition hold open the possibility of policy revision and institutional reform and so set limits—never absolute—on the extent of particular disasters.

Democracy, then, is not simply the recognition of the people as a group that has to be appeased—with subsidies, say, or welfare payments. Nor is it the recognition of each and every citizen as an individual entitled to respect and equal treatment. Welfare payments are important, and equal treatment even more so, but by themselves they are inadequate because they suggest the activity of small groups, benevolent elites, who do the paying and the treating. Democracy is the activity of the rest of us, the rule of the people in their assemblies and committees, arguing over every aspect of the common life. Hence democracy and socialism are, roughly speaking, the same thing: two forms of procedural justice, focused on a certain conception of human *doing* that expresses the deepest values we associate with human *being*. I am sure that these values can be expressed in other ways; sometimes, in the city of Mahagonny, sheer crankiness does the job. They are unlikely to be acted out, however, except within political structures that invite the action, foster it and enclose it, give it shape and force.

We need another city, Mahagonny's opposite. And though

I can't describe it in detail or give a theoretical account of the social forces that might come together to create it, I have a picture of it in my mind. It is not unlike the "great city" of Walt Whitman:[3]

> Where the city stands with the brawniest breed of
> orators and bards;
> Where the city stands that is beloved by these, and
> loves them in return, and understands them;
> Where no monuments exist to heroes, but in the
> common words and deeds;
> Where thrift is in its place, and prudence is in its
> place;
> Where the men and women think lightly of the
> laws;
> Where the slave ceases, and the master of slaves
> ceases;
> Where the populace rise at once against the never-
> ending audacity of elected persons;
> Where fierce men and women pour forth, as the sea
> to the whistle of death pours its sweeping and
> unript waves;
> Where outside authority enters always after the
> precedence of inside authority;
> Where the citizen is always the head and ideal—
> and President, Mayor, Governor, and what not,
> are agents for pay;
> Where children are taught to be laws to them-
> selves, and to depend on themselves;
> Where equanimity is illustrated in affairs;
> Where speculations on the Soul are encouraged;
> Where women walk in public processions in the
> streets, the same as the men,
> Where they enter the public assembly and take
> places the same as the men;

[3] "Song of the Broad-Axe," in *Leaves of Grass*.

Where the city of faithfulest friends stands;
Where the city of the cleanliness of the sexes stands;
Where the city of the healthiest fathers stands;
Where the city of the best-bodied mothers stands,
There the great city stands.

Too healthy, perhaps, for the modern sensibility, too easy in its democratic faith, untouched by the terrors of the twentieth century: and yet I would like, just once or twice, to walk in those processions.

PART I

Liberalism in Retreat

1

Dissatisfaction in the Welfare State

I

One day, not soon, the welfare state will extend its benefits to all those men and women who are at present its occasional victims, its nominal or partial members. That day will not be the end of political history. But it will represent the end of a particular history, and one in which socialists have been very much involved, if not always on our own terms.

It is worth reflecting on what that day will be like—what will we want *then*?—even while we fight to perfect the system of benefits and argue among ourselves about the best strategies. For we are not entering, we are not going to enter, the new world of state-administered prosperity all at once. It is in the nature of the welfare state, I think, that men and women break into it in groups, some sooner, some much later, some with only moderate difficulty, some after long and bloody struggles. Many of us are inside already, better served by machines and bureaucrats than our ancestors ever were by servants and slaves. What do we want *now*?

tic" into a machine, the instrument of its citizens (rather than their mythical common life), devoted to what Jeremy Bentham called "welfare-production." It is judged, as it ought to be, by the amounts of welfare it produces and by the justice and efficiency of its distributive system.

Political unreason survives, of course, and especially in the form of an extraordinary devotion to the modern nation-state and to its leaders, a collective zeal all too often unmitigated by individual interest or by any demand for functional transparency. But here too the direction of political utilitarianism is clearly marked out. Thus an eighteenth-century *philosophe*: "What is patriotism? It is an enlightened love of ourselves, which teaches us to love the government which protects us . . . the society which works for our happiness."[1] This definition suggests that many of us are patriotic, if we are, for wrong or inadequate reasons. I will try to describe some of these reasons a little later on.

Second, the expansion of welfare production gives to the state a new and thoroughly rational legitimacy. The state is always immoral when viewed from the standpoint of its invisible and degraded members. Whatever the ideologies of which they are the primary victims, oppressed classes come eventually to regard the claims of their rulers with a deep-rooted skepticism and hostility. But the claims of the liberals are of a different sort, not mysterious but hypocritical. Therefore the hostility of the oppressed takes a new form: not sullen and inarticulate disbelief, but a positive demand that the claims be realized. Do the middle classes claim to increase the general prosperity? Let them increase *our* prosperity. Do the police claim to defend public security? Let them defend *our* security. Do the rulers of the welfare state claim to maximize the happiness of the greatest number? Let them maximize *our* happiness.

[handwritten margin note: priority of oppressed]

---

[1] Baron d'Holbach, *La Politique Naturelle, ou discours sur les vrais principes du gouvernement* (London, 1773), vol. 2, p. 94.

Now insofar as the state becomes a general welfare state, excluding nobody, it meets these demands and so generates a legitimacy such as no previous political system has ever achieved. If no one is invisible, the state is not immoral. The recognition of its members as concrete individuals with needs and desires may seem a minimal requirement of any political system and hardly capable of producing significant moral attachments. In fact, however, such recognition, when it is finally achieved, will be the outcome of centuries of struggle; the right to be visible is always hard-won, and the liberal state that finally recognizes all men and women and grants them their humanity will inherit from those centuries an extraordinary moral power. The state will never again be so easy to challenge as it was in the days of mass invisibility.

Third, the development of the welfare state has gone hand-in-hand with a transformation in the scale of political organization. This is due not only to increases in the rates of infant and adult survival—the first benefit of the welfare state is life—but also to the progressive extension of political membership to previously invisible people. The tiny political public of an earlier period has been broken into by successive waves of lower-class invaders. It has expanded to absorb each wave; it will probably expand to absorb each future wave.

Liberal theorists and politicians have discovered that there are no necessary limits on the size of the public—so long as its members are conceived as individual recipients of benefits, so long as the problems of political communion, the sharing of a common life, are carefully avoided. Now that there really are concrete benefits to be divided, the first political problem is distribution. And the members of the state, precisely because they are recognized as needful persons, seen by the impersonal public eye and assisted by an impersonal administration, need no longer be able to recognize and assist one another. The size of the citizen body of a Greek polis, like that of an early modern aristocracy, was limited by the requirement that its members be known to each other and so distinguishable from the faceless mass. But once invisibility is banished, the need for political "friendship" is also banished. The members of the welfare state need not have even the most remote acquaintance with each other. And so the welfare state is potentially of infinite extent.

The fourth tendency of successful welfare production is to decrease the importance of politics itself and to turn the state from a political order into an administrative agency. This was always the goal of liberalism, and it is the key reason for the liberal insistence upon the transparent material purposes of the state. Beyond welfare the liberal state cannot go: the world of the mind, philosophy, art, literature, and religion; the world of the emotions, friendship, sex, and love—all these have been set outside its limits. They have been freed from politics, protected against heavy-handed and intolerant magistrates. Simultaneously, liberal politics has been freed from philosophy, art, literature, religion, friendship, sex, and love.

Politics is now concerned only with the provision of a plentiful and enjoyable external world and with the promotion of longevity so that this world can be enjoyed as long as possible. And it has been the assumption of liberal theorists ever since Hobbes and Locke that once security and welfare were assured, once the utilitarian purposes of politics were achieved, men and women would turn away from public to private life, to business and family, or to religion and self-cultivation. Indeed, it was this turning away—which might be called legitimate apathy since it rests on the satisfaction of all recognized needs and desires—that would assure the stability of the liberal achievement. Conflict would disappear; the state would become a neutral agency for the administration of security and welfare. This was a liberal vision even before it was a Marxian one, as Marx himself suggested when he wrote that political emancipation, as practiced by the liberals, "was at the same time the emancipation of civil society from politics."

The state is an instrument and not an end in itself. Politics is an activity with a purpose and not itself an enjoyable activity. These are axioms of liberal enlightenment; the attack upon the opacity of the traditional polity turns out to be an attack upon the value of political life. Why should we be active in politics? asked Thomas Hobbes, and his sarcastic reply suggests the central animus of liberal theory:[2]

—to have our wisdom undervalued before our own faces; by an uncertain trial of a little vain-glory to undergo most certain enmities . . . to hate and be hated . . . to lay open our secret councils and advices to all, to no purpose and without any benefit; to neglect the affairs of our own families.

So long as the state establishes peace and—added later liberal writers—promotes welfare, public activity is a waste of time, positively dysfunctional in the economy of private life. Happiness begins and ends at home. One of the ways in which the welfare state promotes happiness is by encouraging people to stay home. Hence the crucial principles of welfare distribution are, first, that benefits ought to be distributed to individuals and, second, that they ought to be designed to enhance private worlds. In the perfected welfare state, enjoyment will always be private; only administration will be public; the policeman and the welfare administrator will be the only public persons.

Now obviously this is no description of our present experience. Never in human history has politics been so important to so many people, never have so many been active in politics, as in the past century and a half. Never before has the state stood at the center of so large a circle of conflict, agitation, and maneuver. Politics has been the crucial means of becoming visible, of winning recognition for mass needs and desires, and of liberating individuals from social and economic oppression. Nor has it been only a means: political activity has also brought the first joyful sense of membership in a com-

---

[2] *De Cive*, chap. 10, para. 9.

munity. It has provided the positive pleasures of self-assertion and mutual recognition, of collective effort and achievement.

Unlike the welfare state itself, the struggle groups that have demanded and won the various benefits the state now provides—the unions, parties, and movements—have been shaped to a human scale; their members have also been colleagues; they have called one another brethren, citizens, comrades. For a brief moment in time they created a communion that was not mysterious or opaque precisely because it was a common creation. As the organizations of the oppressed win their battles, however, they are gradually integrated into the system of welfare administration. Their purposes are not given up, or not wholly given up, but rather give rise, under new circumstances, to new organizational forms: the struggle groups become pressure groups. Public life ceases to engage the minds and emotions of their members; local activity drops off; popular participation declines sharply. The tenacious sense of detail peculiar to highly qualified bureaucrats replaces the enthusiasm of members: it is more useful, even to the members themselves.

The pleasures of political struggle cannot be sustained once victory has been won. And it is in the nature of the infinitely expandable welfare state that victories can, in fact, be won. Thus it happens that communion is replaced by distribution, generalized aspiration by concrete expectation. Erstwhile militants are isolated and immobilized by the sheer size of the state into which they have won admission, mollified by its apparent legitimacy, by the obvious sincerity of its administrators and the transparency of their purposes. The history of the welfare state begins with the coerced passivity of invisible and degraded men and women, mystified by ideology. And it ends, or will one day end, with the voluntary passivity of enlightened men and women, their human desires recognized and (in part) gratified by the public authorities. So, at any rate, we have been led to believe by liberal writers.

What more can we possibly ask?

III

All these developments—the growing rationality and legitimacy of the state, the vast increase in its size, and the decline of political life—are not only compatible with classical liberal theory, but actually represent its fulfillment. But there is one further corollary of welfare production that raises serious problems for liberalism: the growth of state power.

There can be no question that the development of welfare programs has involved (or required) an extraordinary expansion of the machinery of everyday state administration and therefore an increase in the degree, intensity, and detail of social control. In part, this increase stems directly from the progressive enlargement of bureaucratic systems and from improvements in the training and discipline of their personnel. But it is also closely related to the very nature of the utilitarian service state and to the character of the political struggles of the past century and a half.[3]

For all that time liberals (and socialists too) have been like that character of Gogol's who "wanted to bring the government into everything, even into his daily quarrels with his wife"—though some of them would have stopped short at the seat of domestic bliss. Everywhere else the agents of government have been invited to roam. This was true even during the brief moment of *laissez faire*, for the restrictions on commerce which were then overcome were largely local and corporate, and only the central government could overcome them. Indeed, the state has been an instrument absolutely necessary to reformers of every sort: it shatters the authority of local and traditional elites; it destroys the old corporations and regulates the new ones; it establishes minimal standards

[3] The single greatest factor in the expansion of state power has, of course, been external war, and it is probably true that the greatest dangers posed by the modern liberal state are not those which its own citizens must face. But I have excluded foreign policy and war from this discussion.

for masses of people whose own organizations, however powerful, cannot do so by themselves; it protects racial and religious minorities. It is, so to speak, the crucial licensing agency of modern society, and today virtually the only one; it accords recognitions, turning oppressed subjects into full-fledged members. And it absorbs the power of every defunct agency as it wins the support of every newly enfranchised member.

Nor is its usefulness at an end. Given the continued creation of new groups and the continued raising of the level of material desire, the perfection of welfare production may well be an asymptotic goal and the state an eternally progressive force. But this does not mean that the character of political struggle will remain unchanged as new groups and new desires replace the old. For from a certain point in time, the new groups will almost certainly cease to have the same communal structure as the old. The deprivations of their members are more likely to be experienced by each individual in his private, state-protected world, experienced simultaneously but not shared. Thus the American black is one kind of invisible man, bound to his fellows in a community of suffering and anger and therefore capable of collective action. The person who drives a dangerous car or breathes polluted air is another kind, largely unaware of the risks he shares with others, only marginally aware of the existence of others, and probably incapable of significant efforts on his own behalf. Precisely because of the privatizing results of the benefits he had already won (his automobile, for example), he now stands alone and helpless in the face of one or another sort of corporate power. He is dependent upon the muckraking of free-lance journalists and academic experts and, much more, upon the benevolence of the state.

This benevolence has its price: the increased power of the benevolent administrators, the increasing control over the recipients of benefits. Perhaps the most impressive feature of modern welfare administration is the sheer variety of its coercive and deterrent instruments. Every newly recognized need, every service received, creates a new dependency and so a new social bond.

"A wife and a child are so many pledges a man gives to the world for his good behavior," wrote Jeremy Bentham.[4] This is true only insofar as the world—economy and state—actually provides or promises a decent living to the wife and child. If it does, the pledge is serious indeed. And the better the living the world promises, the more good behavior it requires. Welfare politics thus has a dialectical pattern: pressure from below for more protection or benefits meets pressure from above for better (more disciplined, or orderly, or sociable) behavior. A balance is struck, breaks down, is struck again. Each new balance is achieved at a higher level of welfare production, includes more people, provides new reasons (and new sanctions) for good behavior. Eventually, every anti-social act is interpreted as a demand for increased benefits. So it is. And so welfare is the obvious and only antidote to delinquency and riot. For who would be unwilling, if actually given the chance, to pay the price of social discipline, orderly conduct, hard work, and public decency for the sake of the pursuit of happiness? Only much later does it turn out that the price and the purchase are very nearly the same thing. Happiness *is* good behavior, and this equation, fervently endorsed by the authorities, is the ultimate sanction.

Like the public recognition of needs, so the recognition of individuals—our hard-won visibility—becomes a source of intensified social control. Never have ordinary citizens been so well-known to the public authorities as in the welfare state. We are all counted, numbered, classified, catalogued, polled, interviewed, watched, and filed away. The IBM card is the very means of our visibility, the guarantee that we are not forgotten among so many millions—even as it is simultaneously a symbol of our bondage to the bureaucratic machine.

[4] *The Principles of Morals and Legislation* (Darien, Conn., 1970), p. 54n.

Invisible men are invisible first of all to the officials of the state, and that is a worse bondage. Precisely because they are not seen as citizens, they are exposed to arbitrary cruelty and neglect. Because they are not numbered, they are always treated *en masse*. Because they are never polled, they are thought to have no opinions. Even their crimes, so long as they injure only one another, are not recorded. When their country goes to war, they are impressed (that is, kidnapped), but not conscripted. Gradually, with the development of the welfare state, all this changes. An extraordinary traffic opens up between the visible and the previously invisible sections of society. Individuals and groups win public recognition, learn good behavior, and march out of the slums and ghettos. At the same time, policemen, census takers, recruiting officers, tax collectors, welfare workers, radical organizers, and sociologists (in roughly that historical order) march in.

In the long run, the two parts of society will merge into one world of absolutely visible men and women (that does not mean a world of equals), known not to each other but to the specialists in such knowledge, not personally but statistically. The universal character of this new knowledge doubtless will protect individuals from magisterial whim and prejudice. That is one of its purposes. But it will also involve a new kind of exposure: to the developing administrative sciences of anticipation and prevention.

"It's the anarchy of poverty/Delights me . . ." wrote William Carlos Williams.[5] He was too easily delighted, or rather, his delight was that of an onlooker and not a participant in the "anarchic" culture of the poor. Few people who are actually poor would share it. But having said that, it is still worth adding: It's the regiment of the contented/That haunts me.

---

[5] "The Poor," in *The Collected Earlier Poems* (New York, 1951), p. 415.

IV

Liberals have not been unaware of the dangers of administrative tyranny. Wherever possible they have sought to avoid even benevolent regimentation by giving those whose welfare is at stake "sovereign" power, that is, by establishing governments representative of everyone who receives benefits. In the past that has generally meant of all property owners, for they have been the most important welfare recipients. It obviously means more today, though just how much more is unclear.

The expansion of the range of state benefits and the extension of the ballot to new social groups have been parallel and related processes. Suffrage is the first badge of membership; it is a means of winning benefits and also, presumably, of determining their character and the nature of their administration. In practice, however, it is something less than this. "Welfare without representation," a liberal politician once said, "is tyranny." That is certainly true. But it is not the case that the only alternative to tyranny is a full-scale democracy.

In theory, of course, the purpose of representative government is to make the mass of people all-powerful. Representatives are to be delegates, asserting popular desires, and then legislators, enacting the popular will. Administrators are to be nothing more than servants of the people, bound absolutely by legislative decree. The quality of security and welfare is thus popularly determined, at only one remove. The government of representatives cannot be made responsible to the people on a day-to-day basis, but its general responsibility can be maintained by periodic elections and, more importantly, by continuous political activity between elections. The ultimate defense against bureaucratic omnicompetence is the self-interested assertiveness of ordinary men and women.

If things have not worked out this way, and do not seem likely to, it is at least partly because liberals never developed a system of democratic activism sufficient to bind administrators to representatives and representatives to constituents. Political parties might have served this purpose, but in the United States, at least, parties have not developed as membership organizations capable of stimulating commitment and action on the grass-roots level. Local politics has never been competitive with business and family. Americans have learned that the enhancement of private life through public welfare really does not require any very rigorous and energetic self-government. It may well require a period of sustained struggle, but once that has been won, continued political participation (beyond occasional voting) seems unnecessary and even uneconomical.

Nor is it obvious that the closely articulated representative system that might make such activity worthwhile is really feasible, given the potentially infinite size and the extraordinary administrative complexity of the modern state. Administration has already outdistanced every other branch of government in the sheer accumulation of resources: competent staff, statistical knowledge, patronage, fiscal controls, regulatory powers, secrecy when that is required, publicity when it is not, and so on. Legislative activity has ceased in virtually every respect to be the central feature of the governmental process. It has been replaced in part by administration itself, in part by bargaining between state bureaucrats and the (non-elected) representatives of a great variety of social constituencies.

Thus a modern worker or farmer is far more usefully represented, his interests more successfully defended, by the Washington-based lobbyist of his union than by his locally-based congressman. This is true even though his congressman is elected in a democratic fashion, while the officers of his union are probably co-opted and the lobbyist appointed. It is virtually a law of political life that power will be imitated, that those who seek benefits will copy the organizational style of those who dispense benefits. Today it is palpably the executive rather than the legislative branch of government that is copied. At least, it is copied by those secondary associations already within the welfare system; outside, other models are still possible. In any case, the electoral process has gradually taken on the character of an outer limit, a form of ultimate popular defense rather than of popular self-government, while the day-to-day visibility of workers and farmers and the legitimacy of their government are both maintained by processes largely, though by no means entirely, independent of democratic elections. The modern welfare state is an example of limited government, but not yet of popular sovereignty.

V

The failure of self-government reveals the fundamental difficulty of liberal utilitarianism. Its standard of utility is the welfare of an individual absolutely free to make his own choices and measure his own happiness. In fact, however, no such individual has ever existed.

Men and women live in groups and always find that they have limited choices and share, without having chosen, social standards. If they are ever to be free to choose new limits and standards, they must do so in some cooperative fashion, arguing among themselves, reaching a common decision. But to do this, to act collectively like the sovereign individual of the utilitarians, they must share political power. Government must be responsive to their concrete wills and not merely (as at present) to their conventionally defined desires. If they do not share power, they inevitably become the prisoners of the established social systems that they invade or into which they are admitted. State recognition of new groups obviously affects the structure of social power and value, but there is very little evidence to suggest that it does so in fundamental ways. It clearly does not do so in the ways anticipated during the long struggles for recognition, that is, it does not open the way to social and economic equality. The welfare state has turned out to be perfectly compatible with inequality. Bureaucratic benevolence even bolsters inequality insofar as it neutralizes the struggle groups, decomposing and privatizing popular willfulness. Fundamental social change would require that the state embody this willfulness, inviting its new members to choose their own limits and invent their own standards. This it does not, perhaps cannot, do.

Instead, welfare administrators function, whether consciously or not, as double agents: serving the material interests of the invaders and upholding at the same time the social system that is being invaded. That is why welfare administration, especially in its more direct forms (social work, for example), tends so generally toward paternalism. The administrators are committed in advance to the common limits and standards, to the established modes of security and welfare. They are knowledgeable about these modes and patronizing toward anyone who is ignorant or uncommitted. But the invaders have burst into a world they never made. They have to be helped, guided, educated in the acceptable forms of aspiration and action. They are, in a word, newly licensed to have needs, but not yet intentions or plans of their own.

The perfected welfare state will bring with it an end to the terrible oppressiveness of poverty and invisibility. Once all people are recognized as members (even if only by a distant and powerful government), the sheer magnitude of state terrorism and economic exploitation, and so of human misery, will be enormously reduced. At the same time, it needs to be said that security and welfare are not open-ended categories whose final character will be determined by the freedmen of the liberal state. The pursuit of private happiness may be endless, but its direction, for most of us, is given. The newer welfare recipients are not and are not likely to become self-determining individuals; they remain subject to the determinations of others, not only in the state, but in society and economy as well. Indeed, the established forms of social and economic (that is, corporate) power are likely to be strengthened just as state power has been strengthened: by the general expansion in scale, by the increase in legitimacy that derives from the admission, however reluctant, of all outsiders, by the universal improvement in everyday social behavior, by the new forms of bureaucratic surveillance and record-keeping. For these same reasons, the individual member is taken into account in a new way. When his rulers claim to serve him, the claim is not a lie; it is his political destiny to receive services. The reception of services brings freedom of a limited sort, but of a sort rare enough to be valuable. The citizen of the welfare state is free (and, in many cases, newly enabled) to pursue happiness within the established social and economic system. He is not free to shape or reshape the system, for he has not seized and, except in minimal ways, he does not share political power.

VI

There is no readily accessible meaning in the frequent assertion that socialism lies "beyond the welfare state." No evidence suggests that socialism represents the next stage of history, or that the full development of welfare-production entails a socialist society. "Beyond," in that hopeful phrase, has neither a historical nor a logical sense.

Although socialist parties and movements have often been in the forefront of the struggle for welfare, and above all for equality in the distribution of benefits, it remains true that socialist theory belongs to a tradition of thought and aspiration not only different from, but also in perennial competition with liberal utilitarianism. Against the utilitarians, socialists have argued that mere private life, however enhanced by state action, cannot sustain a significant human culture; that the family does not by itself provide an adequate arena for the human emotions; that man has both a mind and a passion for society; that he requires an active public life. Against the private individual of liberal theory, socialists have defended the free citizen.[6]

This disagreement is in no sense adequately summed up in the words "individualism" and "collectivism." In both traditions, the individual is recognized as the ultimate value; socialists and liberals unite in opposing any ideology that assigns to the state a moral being independent of the willfulness and rationality of its particular members. But to deny the claims of statist ideologies is not necessarily to assert a purely instrumentalist political theory. Individual men and women can still recognize the pleasures of politics, can still choose political life as an end-in-itself. For politics is something more than welfare production. It is a vital and exciting world of work and struggle; of aspiration, initiative, intrigue, and argument; of collective effort, mutual recognition, and *amour social*; of organized hostility; of public venture and social achievement; of personal triumph and failure.

The welfare state offers no satisfactory substitute for any of these. Its theorists claim that all the intellectual and emotional energies of politically active men and women can be rechanneled into private life, and their creativity co-opted by intelligent administrators. Neither of these claims is true. The welfare state requires the virtual withering away of political energy and the disappearance, at least from public life, of any very significant popular creativity. This requirement is first of

[6] Utilitarianism is not, of course, the only form of liberal politics; liberals have also been interested in local government and voluntary association; pluralism, before it became an ideological catchword, was as much a liberal as a socialist theory, and for both liberals and socialists it emphasized the values of political participation and communal life. But I don't think it is unfair to suggest that utilitarianism has been for some time the dominant form of liberalism and, more important, that it has been the central creed of liberals-in-power. I suppose utilitarianism has also been the central creed of socialists-in-power, but socialists, perhaps fortunately for socialism, have been in power less often.

all an extreme restriction upon the pursuit of happiness—because political activity is or might be one of the forms that pursuit takes for many people. Even more important, it involves a surrender of everyone's say in the determination of further restrictions (or expansions), a surrender of any popular role in determining the shape and substance, the day-to-day quality, of our common life. This is the socialist indictment of liberal utilitarianism.

The terms of the indictment have not often been made clear in the recent past because so many of us who regard ourselves as socialists have found a kind of political fulfillment in the struggles for the welfare state. We have allied ourselves with the crowds of people battering at the gates of American society. Their cause has been our public passion. This will continue to be so for some time to come, probably for the foreseeable future. Nevertheless, as we campaign for this or that welfare measure, we are driven to ask ourselves John Stuart Mill's famous question: "Suppose that . . . all the changes in institutions and opinions which you [are] looking forward to could be completely effected at this very instant: would this be a joy and a happiness to you?" And as with Mill, so with us, "an irrepressible self-consciousness distinctly answers, No!"[7]

The welfare state is not the name of our desire. And what is more, the achievement of the welfare state might well entail the end of that public activity which has until now been a joy and a happiness to us. When contemporary writers talk about the "end of ideology," the disappearance of generalized aspiration, they are describing just this closing down of the possibilities for public intellectual and emotional commitment. Though their announcements are at the very least premature, they linger in our minds as disturbing predictions.

[7] *Autobiography*, ed. John Jacob Coss (New York, 1924), p. 94.

VII

But if we are right in thinking that there are human desires that the welfare state cannot fulfill, then surely these desires will continue to be expressed in the form of collective demands and claims. If men and women really do seek the common achievements and shared excitements of politics, then they will have them, in one form or another. And they will produce new ideologies to validate their new activities. In a sense, the welfare state makes these new activities possible, even though it decrees that they must be pursued without the goad of material need and even though its officials hope that they will not in fact be pursued. Political life can now be chosen for its own sake, for the sake of decision making and communal control. I am going to assume that liberal ideologists' predictions of the "end of ideology" are wrong and that large numbers of people (not all people) are prone to make this choice. They demand or will demand some form of political participation, some share in political power. How shall their demand be met?

It might be met by the state itself. For there is no obvious reason why the officials of the welfare state, pressed by their constituents or by their own ambitions, should not move beyond the narrow limits of liberal utilitarianism. They will be, they are already, tempted to do so, for two not entirely consistent reasons: first, to avoid the great difficulties that will inevitably be encountered in the expansion of welfare production beyond its present limits; second, to avoid the boredom of success. In the long run, the second is the more important: while the building of the welfare state will have its exciting and morally significant moments, its administration will not. Its administrators will rarely feel themselves buoyed up and sustained by the zeal of their clients. The pride they may well take in the material services they render will rarely be elevated by the inner conviction of a higher purpose.

But beyond welfare there are many areas in which such purposes might be sought: education, culture, communications, mental health, city planning (and, most tempting and dangerous of all, though not within the range of this discussion, foreign policy). In all these areas the state can be active, in all of them political energies can be expended and emotional commitments made, in all of them common goods can be discovered, goods to be shared and not merely distributed. Imagine, then, that the state moves into the field of "culture-promotion." Won't the possibilities of cooperative choice and political self-determination be enhanced? Won't the citizens of such a state, the recipients not only of material but also of spiritual benefits, feel themselves to be members of a moral community, a world of rights and obligations?

Perhaps they will; our feelings are not always under our own control. Indeed, we know they will, for the modern state always intrudes, in greater or lesser degree, into areas that lie beyond welfare, most often in the name of political socialization. Obviously, it does so with considerable success. But all this has nothing to do with socialism or with a meaningful common life, for reasons that go to the very heart of the theory of citizenship and participation.

Unlike the defenders of the welfare state, theorists of citizenship have always been concerned with the problem of social scale. If human emotional and intellectual needs are to be fulfilled (partially) within political society, they have argued, then that society cannot be of any size or shape. It must be built on a human scale, accessible to our minds and feelings, responsive to our decisions.

Exactly what constitutes a human scale is and ought to be a subject of debate, but this is a debate likely to be carried on chiefly among radical democrats and socialists; it is not a debate in which liberal utilitarians take much interest. For it cannot be established that security and welfare are more efficiently administered to 2 million or 50 million or 200 million people. In fact, it is virtually certain that the quality of security and welfare need not change with the size of the population.

But this is probably not the case with regard to the fulfillment of nonmaterial desires. The quality and authenticity of emotional commitment, for example, do appear to vary with population size, though not absolutely or without reference to other factors: human emotions are more easily manipulated the wider their focus and the more they are cut away from immediate personal experience. Participation in cultural life probably varies the same way: the larger the audience, the more passive its members, the more stereotyped the products they consume. Once again, the formula is too pat, but surely contains an element of truth, and its significance may plausibly be extended to politics as well.

The increasing size of the state, the growing power of administration, the decline of political life: all these turn politics from a concrete activity into what Marx once called the fantasy of everyday life. The state becomes an arena in which men and women do not act but watch the action, and, like other audiences, are acted upon. Patriotic communion is always a fraud when it is nothing more than the communion of an audience with its favorite actors, of passive subjects and heroes of the stage. Our emotions are merely tricked by parades and pageants, the rise and fall of political gladiators, the deaths of beloved chiefs, the somber or startling rites of a debased religion. It could be done to anyone, whereas patriotism ought to be the pride of a particular person, the enjoyment of particular activities.

When the modern state moves beyond welfare, it does not bring us the satisfactions of citizenship, but only vicarious participation, the illusion of a common life. We find ourselves, as if in a dream, living once again in a world that is morally dense and opaque, mystified by ideologies, dominated by leaders whose purposes are not obvious. We are oppressed in the name of a public interest, a national purpose, a solemn commitment, which is neither yours, nor mine, nor ours, in any usual sense of those perfectly simple pronouns. It is difficult not to conclude, as the liberals do, that with the provision of individual material needs, the state reaches or ought to reach its limits. That is the end of its history, the culmination of its legitimacy. There is no state beyond the welfare state.

VIII

The struggle to control the modern state is a struggle for the perfection of the welfare system. Any political leader who claims that it is more than this, who claims, for example, that citizens should do more for the state than the state does for them, is a dangerous man. He aims to avoid the problems of welfare production, or he seeks some sort of totalitarian "transcendence" (or he is preparing the nation for one or another kind of imperial adventure).

The fight over welfare is important enough. Given the most immediate desires of the poor themselves, given the sheer avariciousness of the rich and the powerful, the fight for some minimal standards of distributive justice takes on all the moral significance that has been attributed to it in the past century and a half. Nevertheless, it is not the only fight; nor ought the state to be the only focus of contemporary political struggle. Even if the welfare state were to be perfected under the best possible conditions and under socialist auspices, the dangers of bureaucratic omnicompetence and popular passivity would not be avoided. Nor could a socialist government by itself create a socialist society. That requires a different kind of politics, not the kind to which we are all so well accustomed, aimed permanently at the state, but a politics of immediate self-government, a politics of (relatively) small groups.

Socialist writers have never had a great deal that was new or interesting to say about the state. Despite vague phrases about its withering away, they seem to presuppose, as they probably must, an efficient and benevolent bureaucracy, hovering, so to speak, in the background, its central offices as far away as possible. The chief concern of the best left-wing theorists has always been with that day-to-day cooperation in productive activity which occupies the foreground of social life, with those "life-giving nucleii," as Simone Weil called them, within which the local, immediate character of work and culture is determined.

Such secondary associations exist, or can exist, *within* the welfare state, but insofar as they are of some human value, they exist in permanent tension with the centralized administrative system necessary to welfare production. It is not the natural tendency even of liberal bureaucracy to encourage the formation of autonomous groups. This is so both because of the individualist bias of the welfare system and because of the perennial efforts of administrators to escape the system's utilitarian limits and meet the demand for meaningful citizenship in their own (fraudulent) fashion. What socialism requires, then, is not that the welfare state be surpassed or transcended, whatever that would mean, but that it be held tightly to its own limits, drained of whatever superfluous moral content and unnecessary political power it has usurped, reduced so far as possible to a transparent administrative shell (overarching, protective, enabling) within which smaller groups can grow and prosper. The state is not going to wither away; it must be hollowed out.

What sorts of groups can fill the shell? Two are of especial interest here.

First, the great functional organizations, labor unions, professional associations, and so on: these are the crucial representative bodies of the present day. Their strength and inclusiveness are the best guarantees we can have of the benevolence of the welfare bureaucracy. Unorganized men and women are unrepresented and unprotected men and women, their claims unheard or but distantly heard at the centers of power. If they are benefited at all, they are subject to the most extreme paternalism. Hence the perfection of the welfare state will require the organization of all possible functional groups, even the group of those who receive direct state assistance. In a society that still needs and uses the poor, poverty is itself a function and no one so desperately requires representation as the individual without an adequate income. But all these organizations, as I have argued, tend to become integrated into the welfare system: for them, success *is* integration. They are then trapped in more or less stable bargaining arrangements with governmental or corporate bureaucracies and forced to discipline their own members. They are agents simultaneously of distributive justice and social control. They are not and probably cannot be expected to become arenas of democratic decision making (even though they will be the occasional focus of democratic revolts).

Second, all the local units of work, education, and culture: cities and towns, factories, union locals, universities, churches, political clubs, neighborhood associations, theater groups, editorial boards, and so on. These might be conceived as overlapping circles of engagement and action, closed circles (though not closed in any coercive sense), whose members face inward at least some of the time and within which resources are contained. Here are the most likely arenas of a democratic politics. The great danger of the perfected welfare state is that all or most of them would be broken open, so that resources leak away, independence is lost, and the members turn outward to face the powerful state, where all the action is, from which all good things come. To some extent this has already happened. But the process is by no means so far advanced as some of the more extreme versions of the theory of "mass society" suggest. Associations and neighborhoods continue to provide important social space for agitation and activity. Indeed, it needs to be said that the advance of the liberal state often transforms traditionalist communities (like the old churches) into new political arenas.

At the same time, however, the same process cuts individuals loose, isolates them from communal ties, drives them into a material and then an emotional dependency on the central authorities. It is in response to the expectant faces of these "liberated" individuals that state administrators proclaim the mysteries of national purpose and decide that they must pursue "excellence," or promote culture, or foster solidarity and moral fortitude, rewarding their eager, needful, and bored constituents with inflated rhetoric and byzantine artifacts, and all too often eliciting from them an irrational and unreflective patriotism.

How do "shared understandings" differ from these "mysteries" of nat'l purpose

Now all such pursuits and promotions lie outside the competence of the state; they belong to a different sphere of activity; they require a smaller scale of organization. To make these points, and to make them stick, is the major purpose of socialist politics in the welfare state. It amounts to saying that what we want *next*, and what we want to share, are the pleasures of power. This demand for local self-determination, since it is made in the face of a state whose power is unprecedented in human history, is sometimes called by the melodramatic but useful name "insurgency."[8]

insurgency

IX

Any member of the welfare state who is willful as well as desirous and who seeks some local space in which to act out his willfulness can be called an insurgent. He insists on his intentions as well as his needs. He seeks to close the circle

---

[8] It is one of the odder features of political life that we characterize the common forms of oppression and domination with euphemisms and use words with the most violent connotations to describe modest acts of individual and popular self-assertion. In an article in *Dissent* some five years ago, I used the term "resistance" to suggest the sorts of activities currently called insurgency by various New Left writers. Their word will do as well, though they are no more guerrilla fighters than I am.

(not every circle) against bureaucratic intrusions. He reargues the old democratic proposition that decisions should be made by those who are most affected by them. He calls into question the omnicompetence of the service state and of all the organizations created in its image.

In schools, factories, and neighborhoods, where social workers pursue their errands of decency, or union officials defend the interests of their members, or provosts and deans plan the educational experiences of the young, insurgency is likely to be a perennial phenomenon. It takes the form of wildcat strikes, welfare unions, student rebellions. Already a sense of professional *esprit* is growing up among those who know, or think they know, how to "handle" such outbursts. They believe that insurgency is a repudiation of services rendered, stupidly self-destructive since the services are so obviously helpful, even if they are often ineptly or impersonally delivered. But insurgency is, or rather ought to be, very different from this. Its participants are not concerned that bureaucrats be sensitive and warm, but that they be reticent and limited, less imposing, less intrusive than they often are. Insurgency is a demand that bureaucratic services make possible, instead of replacing, local decision making. It is a denial of conventional definitions of good behavior and an effort to make the "helpfulness" of the welfare bureaucracy into the starting point of a new politics of self-government.

In the long run, the issue for socialists is not state power, but power *right here*, on this shop floor, in this faculty or university, in this city. And the central assumption of insurgent politics is that such power must always be won "from below" —which is also to say, against all the odds. For the triumph of benevolent bureaucrats in virtually every local society has been one of the results, not so much inevitable as overdetermined, of the entrance of large numbers of men and women into the political world. It is a function of integration. Most of the previous forms of radical politics have involved demands for wider and wider systems of integration; every successful revolution has produced such a system. But insurgency is different from revolution (more limited, more immediate) precisely in that it seeks no more extensive unity, but calls instead for the multiplication of diverse and independent unities. And it begins this process with a modest but urgent demand for a share *right now* in the management of *this* community.

Whether this can actually be won and the victory sustained is another matter. It is not difficult to imagine a kind of permanent insurgency, generating marginal but never major disturbances in the welfare state, always asserting its claims, never able to enforce them, capable of staging riots, never capable of building a movement or a new community. Newspapers provide us daily with intimations of such a pattern. We don't know if local organizations of rebellious citizens can displace entrenched officials, sanctioned and supported by the central state; nor do we know if they can win any substantial allegiance from their own members, establish some more or less effective control over the local politics of work and culture, and generate significant cooperative activity. Above all, we don't know if they can create new patterns of democratic responsibility, so that the militants of this or that rebellion don't simply become a new elite. That success is possible must be the socialist's faith, or better, the wager that sustains his commitment.

The politics of insurgency and the politics of welfare obviously overlap, both in time and technique, and their different purposes are sometimes confused. Insurgency has been a prominent feature of every struggle for recognition: the sit-in strikes of the 1930s are the classic example. Oppressed groups must always win enough power either to threaten those who refuse to grant their demands or to threaten the general peace and so compel state intervention on their behalf. And "enough" power always means power in some particular place, sufficient to inflict some particular injury. But so long as the goal of the oppressed is (as it ought to be) membership in the greater society rather than autonomy for the smaller one,

But what if an insurgent group is always potentially at odds w/ the state, at least at particular points? What if there are (potentially) permanent conflicts bet. the state & a secondary assoc.?

50

such power is largely a means to an end, and it disappears when it ceases to be a necessary means. It gradually seeped away from local unions, for example, once bargaining rights were granted at the national or industry-wide level—and while this was certainly a democratic loss, it must be admitted that what the workers wanted could only have been won at the national level.

Something like this will probably happen in the civil rights movement also, since the problems of the black community cannot be solved until its activists transform such local power as they may win into national recognition and full admission into the welfare state. And this suggests a lesson that is not easy to learn: political power must always be twice-won. It must be won first with the help of the state or through the creation of parallel bureaucracies against established local or corporate elites. Then it must be won again by new popular forces against the state. In the United States today, the blacks are still fighting the first battle; Americans who have already won national recognition are, one hopes, ready for the second.

Socialism, it has often been said, requires decentralization. But that is not quite accurate, for it implies a process that begins at the center and is, just for that reason, inherently unlikely. It suggests that socialism awaits the triumph of a national movement whose leaders are ready to sponsor the fragmentation and dispersion of the power they have so recently won. Once installed, however, won't they be driven to realize how much good might be done with their power if only it is kept intact . . . for a few years more? (And they may well be right.) We ought, of course, to insist that state officials do as much as they can to encourage the growth of secondary associations independent of themselves, not forming wheels within wheels of their welfare-producing machine. But the vitality of such associations depends finally on those who associate.

Socialism then requires insurgency, that is, self-government within the welfare state and against it whenever necessary. And it is the great paradox of socialist politics that the state toward which we must always remain tense, watchful, and resistant is or will almost certainly become the most legitimate, rationally purposive, and powerful state that has ever existed.

(1980)

Postscript

In the years since I wrote this essay, a number of writers have argued that a "legitimation crisis" exists in advanced capitalist societies. The argument, especially in Habermas's version, is complex and sometimes difficult to follow. I cannot engage it here. But I do not believe that the delegitimation thesis has been successfully defended in the case of the democratic welfare state. It may be true that welfare politicians promise more than they can deliver and so undermine their own authority. Or it may be that the people see more clearly what the state is for, what it ought to be doing, and judge more harshly the performance of its officials. Or it may be that the welfare system undermines local communities and weakens the structures that make cooperation and self-restraint possible. And in any of these cases, or in all of them, the welfare state is harder to manage than older regimes; it is more difficult for politicians-in-power to hold on to their power and to win re-election. But this is only to say that there are more concrete and publicly acceptable reasons for complaint in the welfare state, more individuals ready to complain, fewer individuals willing to defer to the powers-that-be.

What is most striking about contemporary politics, however, is that there is so little opposition to the welfare state as a whole. There is no serious revolutionary program for dismantling it or for replacing it with some radically different institutional arrangement. The planning, welfare, and regulatory functions of the state are widely accepted today, and no serious writer has argued that these functions can be carried out in the absence of political authority and administrative machinery of roughly the sort we already have. Proposals for decentralization and workers' control are of the "hollowing out," not the "withering away" variety. They leave the framework intact because their authors have no alternative to it, no other means for the coordination of economic activity, the protection of the environment, the provision of basic services. Certainly, it can safely be said of the Left that whatever else its activists propose to do, they don't propose to dispense with state action in any of these areas. Was there ever a period in modern (post-medieval) history, before this one, when no significant political group imagined itself ready to transform the state? Republicanism, democracy, proletarian dictatorship, anarchy: these were alternative visions. Today, the boldest alternatives, and the most interesting ones, focus on lower levels of organization and authority, smaller units within which, perhaps, more can be achieved.

I don't think by any means that we are or will be free of crisis. But it is hard to imagine what political earthquake could shake the structure of welfare democracy . . . and throw up something better. And it is dangerous to imagine that, were such an earthquake to occur, the hopes of leftists for internal transformation—local initiative, communal autonomy —would be advanced.

(1980)

2

Civility and Civic Virtue in Contemporary America

I

Decline and fall is the most common historical perception, even among intellectuals who are happiest when they have uncommon ideas. I want to examine this perception in its most important contemporary form, which is also a recurrent form. "We have physicists, geometers, chemists, astronomers, poets, musicians, painters," wrote Rousseau in 1750, "we no longer have citizens. . . ."[1] Here in the United States we still do have

[1] Jean Jacques Rousseau, "Discourse on the Sciences and Arts," in *The First and Second Discourses*, ed. Roger D. Masters (New York, 1964), p. 59.

citizens, but it is frequently said of them (by old liberals and new conservatives) that their commitment to the political community is less profound than it once was, that there has been a decline of civic virtue and even of ordinary civility, an erosion of the moral and political qualities that make a good citizen. It is hard to know how to judge statements of this kind. They suggest comparisons without specifying any historical reference point. They seem to be prompted by a variety of tendencies and events that are by no means uniform in character or necessarily connected: the extent of draft resistance during the Vietnam War, the domestic violence of the middle and late 1960s, the recent challenges to academic freedom, the new acceptance of pornography, the decline in the fervor with which national holidays are celebrated, and so on.

Perhaps one way of judging these phenomena is to ask what it is we expect of citizens—of citizens in general but also of American citizens in particular, members of a liberal democracy, each of whom represents, as Rousseau would have said, only 1/200,000,000th of the general will. What do we expect of one another? I am going to suggest a list of common expectations; I shall try to make it an exhaustive list. Working our way through it, we shall see that we are the citizens we ought to be, given the social and political order in which we live. And if critics of our citizenship remain dissatisfied, then it will be time to ask how that order might be changed.

1. We expect some degree of commitment or loyalty—but to what? Not to *la patrie*, the fatherland: that concept has never captured the American imagination, probably because, until very recently, so many of us were fathered in other lands. Not to the nation: the appearance of an American nationality was for a long time the goal of our various immigrant absorption systems, but this goal has stood in some tension with the practical (and now with the ideological) pluralism of our society. Most of those who mourn our lost civility would not, I think, be happy with an American nationalism. Not to the state,

in a temper to submit to conscription. The people have too fresh and strong a feeling of the blessings of civil liberty to be willing thus to surrender it. . . . Laws, Sir, of this nature can create nothing but opposition. A military force cannot be raised, in this manner, but by the means of a military force. If the administration has found that it cannot form an army without conscription, it will find, if it ventures on these experiments, that it cannot enforce conscription without an army.

We have come a long way since those days, a way marked as much by changes in the external world as in our domestic society. The domestic changes have been made only gradually and, as Webster predicted, in the face of constant opposition. It is worth remembering how recent a creation the docile draftee is before we mourn his disappearance (has he disappeared?) as a loss of American virtue. In 1863, the first conscription law was fiercely resisted—over one thousand people died in the New York draft riot of that year—and it was massively evaded during the remainder of the Civil War. The draft was still being evaded on a large scale in World War I, particularly in rural areas where it was easy to hide. And who can say that the young men who took to the woods in 1917 were not reaffirming the values of an earlier America? They would have grabbed their rifles readily enough had the Boche marched into Kentucky. Perhaps that is the only true test of their citizenship.

The citizen-soldier defends his hearth and home, and he also defends the political community within which the enjoyment of hearth and home is made possible. His fervor is heightened when that community is in danger. Armies of citizens, like those of Rome or the first French Republic or Israel today, are born in moments of extreme peril. Once the peril abates, the fervor declines. The armies of great powers must be sustained on a different basis, and the long-term considerations that lead them to fight here or there, in other people's countries, when there is no immediate or visible threat to their own, can hardly be expected to evoke among their citizens a passionate sense of duty. Perhaps these citizens have an obligation to fight, in obedience, say, to laws democratically enacted, but this is not the same obligation that American publicists meant to stress when they made the minuteman a mythic figure. It has more to do with law-abidance than with civic courage or dedication.

3. We expect citizens to obey the law and to maintain a certain decorum of behavior—a decorum that is commonly called civility. That word once had to do more directly with the political virtues of citizenship: one of its obsolescent meanings is "civil righteousness." But it has come increasingly to denote only social virtues; orderliness, politeness, seemliness are the synonyms the dictionary suggests, and these terms, though it is no doubt desirable that they describe our public life, orient us decisively toward the private realm. Perhaps this shift in meaning is a sign of our declining dedication to republican values, but it actually occurred some time ago and does not reflect on ourselves and our contemporaries. For some time, Americans have thought that *good behavior* is what we could rightly expect from a citizen, and the crucial form of good behavior is everyday law-abidance. Has this expectation been disappointed? Certainly many people write as if it has been. I am inclined to think them wrong, though not for reasons that have much to do with republican citizenship.

If we could measure the rate and intensity of obedience to law—not merely the nonviolation of the penal code, but the interest, the concern, the anxiety with which citizens *aim* at obedience—I am certain we would chart a fairly steady upward movement in every modernizing country, at least after the initial crisis of modernization is past. Contemporary societies require and sustain a very intense form of social discipline, and this discipline is probably more pervasive and more successfully internalized than was that of peasant societies or of small towns and villages. We have only to think about our own lives to realize the extent of our submission to what

Max Weber called "rational-legal authority." It is reflected in our time sense, our ability to work methodically, our acceptance of bureaucratic hierarchies, our habitual orientation to rules and regulations. Consider, for example, the simple but surprising fact that each of us will, before next April 15, carefully fill out a government form detailing our incomes and calculating the tax we owe the United States—which we will then promptly pay. The medieval tithe, if it was ever a realistic tax, was socially enforced; our own tax is individually enforced. We ourselves are the calculators and the collectors; the tax system could not succeed without our conscientiousness; the police could never cope with massive evasion. Surely the American income tax is a triumph of civilization. There are very few political orders within which one can imagine such a system working; I doubt that it would have worked, for example, in Tocqueville's America.

But I want to turn to two other examples of our relative civility which speak more directly to the concerns of our recent past, which have to do, that is, with violence. In 1901, David Brewer, an associate justice of the U.S. Supreme Court, delivered a series of lectures on American citizenship at Yale University, in the course of which he worried at some length about the prevalence of vigilante justice and lynch law.[4] This was the peculiarly American way of "not tarrying for the magistrate." "It may almost be regarded," Brewer said, "as a habit of the American people." Clearly, our habits have changed; in this respect, at least, we have grown more law-abiding since the turn of the century. The police sometimes take the law into their own hands, but they are our only vigilantes; ordinary citizens rarely act in the old American way. This is not the result of a more highly developed civic consciousness, but it is a matter of improved social discipline, and it also suggests that, despite our popular culture, we are less ready for violence, less accustomed to violence, than were earlier generations of American citizens.

[4] David Brewer, *American Citizenship* (New York, 1902), pp. 102 ff.

We are also less given to riot; if nineteenth-century statistics are at all reliable, our mobs are less dangerous to human life. The most striking thing about the urban riots of the 1960s—apart from the surprise which greeted them, for which our history offers no justification—is the relatively small number of people who were actually killed in their course. By all accounts, riots were once much bloodier: I have already mentioned New York's "bloody week" of 1863. They also seem to have been more exuberantly tumultuous, and the tumult more accepted in the life of the time. Here, for example, are a set of newspaper headlines from New York in 1834:[5]

A Bloody Fight
Mayor and Officers Wounded
Mob Triumphant
The Streets Blocked by Fifteen Thousand Enraged Whigs
Military Called Out

These lines describe an election riot, not uncommon in an age when party loyalties were considerably more intense than they are now and a far higher proportion of the eligible voters were likely to turn out on election day. The accompanying news story does not suggest that the rioters or their leaders were extremists or revolutionaries. They apparently were ordinary citizens. Our own riots were also the work of ordinary citizens, though not of the contemporary equivalents of Whigs, Orangemen, or even Know-Nothings. They seem to have been peculiarly disorganized, each of them less a communal event than a series of simultaneous acts of in-dividual desperation. They were more frightening than the earlier riots, because the participants lacked any common memberships or shared *esprit,* and also less dangerous. Per-haps this change is appropriate to a liberal society: if civility is restrained and privatized, then so must incivility be.

This last example suggests a certain tension between civility

---

[5] Quoted in Cunliffe, *Soldiers and Civilians,* p. 93.

and republican citizenship. Indeed, in the early modern period, one of the chief arguments against republicanism was that it made for disorder and tumult. Faction fights, party intrigues, street wars, instability, and sedition: these were the natural forms of political life in what Thomas Hobbes called "the Greek and Roman anarchies," and so it would be, he argued, in any similar regime.[6] He may have been right, in some limited sense at least. The improvement in social discipline seems to have been accompanied by a decline in political passion, in the lively sense of public involvement that presumably characterized the enraged Whigs of 1834 and other early Americans. I shall have more to say about this when I turn to the general issue of political participation. But first it is necessary to take up another aspect of our new civility.

4. We expect citizens to be tolerant of one another. This is probably as close as we can come to that "friendship" which Aristotle thought should characterize relations among members of the same political community. For friendship is only possible within a relatively small and homogeneous city, but toleration reaches out infinitely. Once certain barriers of feeling and belief have been broken down, it is as easy to tolerate five million people as to tolerate five. Hence toleration is a crucial form of civility in all modern societies and especially in our own. But it is not easily achieved. Much of the violence of American history has been the work of men and women resisting its advance in the name of one or another form of local and particularized friendship or in the name of those systems of hierarchy and segregation that served in the past to make pluralism possible. It's probably fair to say that resistance has grown weaker in recent years; the United States is a more tolerant society today than at any earlier period of its history. Of course, we need to be more tolerant; it's as if, once we commit ourselves to toleration, the demand for it escalates; it is no longer a question of a recognized range of religious and

---

[6] *De Cive*, chap. 12, para. 3.

political dissidence, but of the margins beyond the range. Even the margins are safer today; more people live there and with less fear of public harassment and social pressure. It is precisely these people, however, who seem to pose a problem for us, who lead us to worry about the future of civic virtue. A curious and revealing fact, for their very existence is a sign of our civility.

The problem is that many Americans who find it easy (more or less) to tolerate racial and religious and even political differences find it very hard to tolerate sexual deviance and countercultural life-styles. One day, perhaps, this difficulty will be remembered only as a passing moment in the painful development of an "open" or "liberated" society. But it doesn't feel that way now; it feels much more drastic, so that intelligent people talk of the end of civilization, all coherence gone, the fulfillment of this or that modernist nightmare. For surely (they rightly say) political society requires and rests upon *some* shared values, a certain spiritual cohesion, however limited in character. And a commitment to moral *laissez faire* does not provide any cohesion at all. It undermines the very basis of a common life, because the ethic of toleration leads us to make our peace with every refusal of commonality. So we drift apart, losing through our very acceptance of one another's differences all sense of kinship and solidarity.

This is undoubtedly overstated, for the fact is that we do co-exist, not only Protestants, Catholics, and Jews; blacks and whites; but also Seventh Day Adventists, Buddhists, and Black Muslims; Birchers and Trotskyites; sexual sectarians of every sort, homo and hetero. Nor is it a small thing that we have made our peace with all these, for the only alternative, if history has any lessons at all, is cruelty and repression. Liberalism may widen our differences as it widens the range of permissible difference, but it also generates a pattern of accommodation that we ought to value. It would be foolish to value it, however, without noticing that, like other forms of civility, this pattern of accommodation is antithetical to political activism. It tends to insulate politics from group conflict, to promote among citizens a general indifference toward the opinions of their fellows, to freeze the intolerant out of public life (they are disproportionately represented, for example, among nonvoters). It stands in the way of the personal transformations and new commitments that might grow out of a more open pattern of strife and contention. It makes for political peace; it makes politics less dangerous and less interesting. And yet our notions about citizenship lead us to demand precisely that citizens *be interested* in politics.

5. We expect citizens to participate actively in political life. Republicanism is a form of collective self-government, and its success requires, at the very least, that large numbers of citizens vote and that smaller numbers join in parties and movements, in meetings and demonstrations. No doubt, such activity is in part self-regarding, but any stable commitment probably has to be based and is in fact usually based on some notion of the public good. It is, then, virtuous activity; interest in public issues and devotion to public causes are the key signs of civic virtue.

Voting is the minimal form of virtuous conduct, but it is also the easiest to measure, and if we take it as a useful index, we can be quite precise in talking about the character of our citizenship. Participation in elections, as Walter Dean Burnham has shown, was very high in the nineteenth century, not only in presidential contests, but also in off-year congressional and even in local elections.[7] Something like four-fifths of the eligible voters commonly went to the polls. "The nineteenth century American political system," writes Burnham, ". . . was incomparably the most thoroughly democratized of any in the world." A sharp decline began around 1896 and continued through the 1920s, when the number of eligible voters actually voting fell to around two-fifths. Rates of participation rose in

[7] Walter Dean Burnham, "The Changing Shape of the American Political Universe," *American Political Science Review* 59 (March 1965): 7–28.

the 1930s, leveled off, rose again in the 1950s, leveled off again—without coming close to the earlier figures. Today, the percentage of American citizens who are consistent nonparticipants is about twice what it was in the 1890s. By this measure, then, we are less virtuous than were nineteenth-century Americans, less committed to the public business.

The reasons for this decline are not easy to sort out. Burnham suggests that it may have something to do with the final consolidation of power by the new industrial elites. The triumph of corporate bureaucracy was hardly conducive to a participatory politics among members of the new working class or among those farmers who had been the backbone of the Populist movement. Some workers turned to socialism (Debs got a million votes in 1912), but far more dropped out of the political system altogether. They became habitual nonvoters, at least until the CIO brought many of the men and women it organized back into electoral politics in the 1930s. If this account is right—and other accounts are possible—then nonvoting can be seen as a rational response to certain sorts of social change. No doubt, it was also functional to the social system as a whole. The decline in participation during a period of increasing heterogeneity and rapid urbanization probably helped stabilize the emerging patterns of law-abidance and toleration. Certainly American society would have been far more turbulent than it was had new immigrants, urban dwellers, and industrial workers been actively involved in politics. That is not to argue that they should not have been involved, only that people who set a high value on civility should not complain about their lack of civic virtue.

A recent study of political acts more "difficult" than voting —giving money, attending meetings, joining organizations— suggests that there was a considerable increase in participation in the course of the 1960s.[8] Not surprisingly, this increase coincided with a period of turmoil and dissension of which

[8] Sidney Verba and Norman H. Nie, *Participation in America: Political Democracy and Social Equality* (New York, 1972), especially chap. 14.

it was probably both cause and effect. One might impartially have watched the events of that time and worried about the loss of civility and rejoiced in the resurgence of civic virtue. The connection between the two is clear enough: people are mobilized for political action, led to commit themselves and to make the sacrifices and take the risks commitment requires, only when significant public issues are enlarged upon by the agitators and organizers of movements and parties and made the occasion for exciting confrontations. These need not be violent confrontations; violence draws spectators more readily than participants. But if the issues are significant, if the conflict is serious, violence always remains a possibility. The only way to avoid the possibility is to avoid significant issues or to make it clear that the democratic political struggle is a charade whose outcome won't affect the resolution of the issues—and then rates of participation will quickly drop off.

The civil rights and antiwar agitations of the 1960s demonstrate that there are still dedicated citizens in the United States. But the activity generated by those movements turned out to be evanescent, leaving behind no organizational residue, no basis for an ongoing participatory politics. Perhaps that is because not enough people committed themselves. The national mood, if one focuses on the silence of the silent majority, is tolerant and passive—in much the same way as it was tolerant and passive in the face of prohibitionists or suffragettes or even socialists and communists in the 1930s: that is, there is no demand for massive repression and there is no major upsurge in political involvement. It is also important, I think, that the two movements of the sixties did not link up in any stable way with either of the established parties. Instead of strengthening party loyalty, they may well have contributed to a further erosion. If that is so, even rates of electoral participation will probably fall in the next few years, for parties are the crucial media of political activism. These two failures—to mobilize mass support, to connect with the established parties—may well suggest the general pattern of political life in America today. For most of our citizens, politics is no vocation. They think it a duty to vote, but they have no deep commitment to a creed or party, and only about half of them bother to vote. Beyond that, they are wrapped up in their private affairs and committed to the orderliness and proprieties, and to the pleasures, of the private realm. Though they are tolerant, up to a point, of political activists, they regard politics as an intrusion and they easily resist the temptations of the arena. This makes life hard for the smaller number of citizens who are intermittently moved by some public issue and who seek to move their fellows. It may help explain the frenetic quality of their zeal and the way some of them drift, in extreme cases, into depression and madness. The institutional structures and the mass commitment necessary to sustain civic virtue simply don't exist in contemporary America.

II

The ideals of citizenship do not today make a coherent whole. The citizen receives, so to speak, inconsistent instruction. Patriotism, civility, toleration, and political activism pull him in different directions. The first and last require a kind of zeal—that is, they require both passion and conviction—and they make for excitement and tumult in public life. It is often said that the worst wars are civil wars because they are fought between brethren. One might say something similar about republican politics: because it rests on a shared commitment, it is often more bitter and divisive than politics in other regimes. Civility and tolerance serve to reduce the tension, but they do so by undercutting the commitment. They encourage people to view their interests as fragmented, diverse, and private; they make for quiet and passive citizens, unwilling to intrude on others or to subject themselves to the discipline of a creed or party. I am not going to argue that

67

we need choose in some absolute way one or the other of these forms of political life. What exists today and what will always exist is some balance between them. But the balance has changed over the years: we are, I think, more civil and less civically virtuous than Americans once were. The new balance is a liberal one, and there can be little doubt that it fits the scale and complexity of modern society and the forms of economic organization developed in the United States in the twentieth century. What has occurred is not a decline and fall but a working out of liberal values—individualism, secularism, toleration—and at the same time an adjustment to the demands of capitalist modernity.

The new citizenship, however, leaves many Americans dissatisfied. Liberalism, even at its most permissive, is a hard politics because it offers so few emotional rewards; the liberal state is not a home for its citizens; it lacks warmth and intimacy. And so contemporary dissatisfaction takes the form of a yearning for political community, passionate affirmation, explicit patriotism. These are dangerous desires, for they cannot readily be met within the world of liberalism. They leave us open to a politics I would find unattractive and even frightening: a willful effort to build social cohesion and political enthusiasm from above, through the use of state power. Imagine a charismatic leader, talking about American values and goals, making war on pornography and sexual deviance (and then on political and social deviance), establishing loyalty oaths and new celebrations, rallying the people for some real or imagined crisis. The prospect is hard to imagine without the crisis, but given that, might it not be genuinely appealing to men and women cut off from a common life, feeling little connection with their neighbors and little connection with the past or future of the republic? It would offer solidarity in a time of danger, and the hard truth about individualism, secularism, and toleration is that they make solidarity very difficult. The recognition of this truth helps explain, I think, the gradual drift of some American intellectuals toward a kind of communal conservatism. Thus, the fulsome debate, some years ago, about our "national purpose" (a liberal nation can have no collective purpose) and the new interest in the possibilities of censorship both suggest the desire to shape citizens in a common mold and to raise the pitch of their virtue.

Even assuming these are the right goals, however, that is the wrong way to reach them. It begins on the wrong side of the balance, with an attack on the heterogeneity of liberal society, and so it poses a threat to all our (different) beliefs, values, and ways of life. I want to suggest that we start on the other side, by expanding the possibilities for a participatory politics. In the liberal world, patriotic feeling and political participation depend on one another, it seems to me, in a special way. For Rousseau and for classical republicans generally, these two rested and could only rest on social, religious, and cultural unity. They were the political expressions of a homogeneous people. One might say that, for them, citizenship was only possible where it was least necessary, where politics was nothing more than the extension into the public arena of a common life that began and was sustained outside. Under such conditions, as John Stuart Mill wrote of the ancient republics and the Swiss cities, patriotism is easy; it is a "passion of spontaneous growth."[9] But today, society, religion, and culture are pluralist in form; there is no common life outside the arena, and there is less and less spontaneous patriotism. The only thing that we can share is the republic itself, the business of government. Only if we actually do share that are we *fellow* citizens. Without that, we are private men and women, radically disjoined, confined to a sphere of existence which, however rich it can be and sometimes is in liberal society, can never satisfy our longing for cooperative endeavor, for *amour social*, for public causes and effects.

Among people like ourselves, a community of patriots

[9] John Stuart Mill, "M. de Tocqueville on Democracy in America," in *The Philosophy of John Stuart Mill*, ed. Marshall Cohen (New York, 1961), p. 158.

would have to be sustained by politics alone. I don't know if such a community is possible. Judged by the theory and practice of the classical republics, its creation certainly seems unlikely: how can a common citizenship develop if there is no other commonality—no ethnic solidarity, no established religion, no unified cultural tradition? When I argued that the contemporary balance of civility and civic virtue is appropriate to a liberal society, I was making the classical case for the connection of society and state, of everyday life and political commitment. I did not mean, however, to make a determinist case. One can always strain at the limits of the appropriate; one can always act inappropriately. And it is not implausible to suggest that social circumstance, like Machiavelli's fortune, is the arbiter of only half of what we do. Given liberal society and culture, certain sorts of dedication may well lie beyond our reach. But that is not to say that we cannot, so to speak, enlarge the time and space within which we live as citizens. This is the working principle of democratic socialism: that politics can be opened up, rates of participation significantly increased, decision making really shared, without a full-scale attack on private life and liberal values, without a religious revival or a cultural revolution. What is necessary is the expansion of the public sphere. I don't mean by that the growth of state power—which will come anyway, for a strong state is the necessary and natural antidote to liberal disintegration—but a new politicizing of the state, a devolution of state power into the hands of ordinary citizens.

Three kinds of expansion are required. They add up to a familiar program which I can only sketch here in the briefest possible way: a radical democratization of corporate government, so that crucial decisions about the shape of the economy are clearly seen to be the public's business; the decentralization of governmental activity so as to alter the scale of political life and increase the numbers of men and women able to play an effective part in everyday decision making; the creation of parties and movements that can operate at different levels of government and claim a greater degree of individual commitment at every level than our present parties can. All this is needed if patriotism is to be nourished, in the absence of social and cultural cohesion, by what Mill calls "artificial means," political arrangements that foster activity and participation.

How patriotism of this sort might actually be achieved, I cannot consider here. It is more important for my present purposes to acknowledge that achieving it (or trying to) will significantly raise the levels of intensity and contention in our politics, and even the levels of intolerance and zeal. Militancy, righteousness, indignation, and hostility are the very stuff of democratic politics. The interventions of the people are not like those of the Holy Ghost. For the people bring with them into the arena all the contradictions of liberal society and culture. And the political arena is in any case a setting for confrontation. Politics (unlike economics) is inherently competitive, and when the competition takes place among large groups of citizens rather than among the king's favorites or rival cliques of oligarchs, it is bound to be more expressive, more feverish, and more tumultuous.

And yet it is only in the arena that we can hope for a solidarity that is spontaneous and free. *E pluribus unum* is an alchemist's promise; out of liberal pluralism no oneness can come. But there is a kind of sharing that is possible even with conflict and perhaps only with it. In the arena, rival politicians have to speak about the common good, even if they simultaneously advance sectional interests. Citizens learn to ask, in addition to their private questions, what the common good really is. In the course of sustained political activity enemies become familiar antagonists, known to be asking the same (contradictory) questions. Men and women who merely tolerated one another's differences recognize that they share a commitment—to *this* arena and to the people in it. Even a divisive election, then, is a ritual of unity, not only because it has a single outcome, but also because it reaffirms the existence of the arena itself, the public thing, and the sovereign people. Politics is a school of loyalty, through which we make the republic our moral possession and come to regard it with a kind of reverence. And election day is the republic's most important celebration. I don't want to exaggerate the awe a citizen feels when he votes, but I do think there is awe, and a sense of pride, at least when the issues being decided are really important and the political order is built to a human scale. There can even be civility in the arena, courtesy, generosity, a concern for rules (especially, as in war, among professionals and veterans)—though one must expect something else much of the time.

In saying all this, I am repeating an argument that Lewis Coser made in the 1950s in his book *The Functions of Social Conflict*. The argument is worth repeating since the conflicts of the 1960s do not seem to have confirmed it for most Americans. Nor would it be irrational to recognize, with Coser, the "integrative functions of antagonistic behavior" and decide, nonetheless, to live with some lesser degree of integration.[10] I am inclined to think that we can have civility and law-abidance without any intensification of patriotism and participation. No doubt, the present balance is unstable, but so is every other; we have to choose the difficulties we shall live with. What we cannot have, and ought not to ask of one another, under present conditions, is civic virtue. For that we must first create a new politics. I have tried to suggest that it must be a socialist and democratic politics and that it must not supersede but stand in constant tension with the liberalism of our society.

(1974)

---

[10] *The Functions of Social Conflict* (Glencoe, Ill., 1956), chap. 7.

3

Watergate without
the President

When Hamlet says, "Something is rotten in the state of Denmark," he means that the king is corrupt. He is not making any comment on the political life of ordinary Danish citizens, for there were no citizens and no political life outside the court at Elsinore. Today, we all know that something is rotten in our own state, but it isn't enough to point to the White House or to Richard Nixon and his entourage. The malaise seems more general. After all, this is a republic; our politics doesn't focus exclusively on a single person. Republics don't rot so readily as monarchies, nor is the rot cut out so easily. We need to talk about Watergate without the president.

Consider first the illegal transfer of vast sums of money from private corporations to the executive branch of the federal government in exchange for favors of one sort or another. Perhaps everyone knew that this kind of thing went on in Washington, but it is astonishing that it went on so *ordinarily,* with special offices set up to handle "government relations" and elaborately devious channels worked out to conceal the actual sources of the cash. What can one say about the financial integrity or civic courage of the American corporate elite? They would appear to possess very little of either. Nor do they seem even remotely equipped to play the role set out for them in American ideology. Conventional doctrine holds that corporate power balances governmental power. But if the corporations are so dependent on the state as to be infinitely open to extortion, where's the balance? Or, if state officials are so dependent on corporate funds as to be infinitely open to bribery, where's the balance?

Yet I can see here the makings of a new *laissez faire*. If the market in power and influence were only let alone, we will soon be told, if nosy reporters and ambitious attorneys general stopped interfering, a delicate equilibrium would establish itself between the government and the corporations—that is, between extortion and bribery—a new kind of harmony, with both sides deterred from asking more than the market would bear. Perhaps it needs to be said that that's not a satisfactory arrangement in a republic, when what is being transacted every time money changes hands is the public business. We expect the public business to be done in public, with reasons and arguments—not cash—exchanged, and with the rest of us watching. It is not easy to guarantee that, however, and clearly the private power of the corporations is not even the beginning of a guarantee; it is instead an inducement to criminal intrigue, to the highest form of white-collar crime.

So socialists have a new opportunity to argue their case: might not the independence of corporate leaders be enhanced if they were public officials, put on their oath, subject to periodic review, responsible to a determinate constituency?

The corporate leaders who have done best in the current crisis are the newspaper owners and television managers—perhaps because they are the most public of our private men; their work is subject to a daily audit. But however grateful we are to them and to the reporters and commentators they employ, the political role of the media ought to worry us. In the 1950s and early 1960s, the burdens of social change were carried by the courts. Today, the burdens of political reform are being carried by the newspapers and the networks. It made us uneasy then to see such large effects worked by judges who had never been elected, who had no representative function. We thought their achievements less secure than they would have been had Congress joined wholeheartedly in the work—and we were right.

But the courts were active precisely because our elected representatives abdicated their moral and political responsibilities. The role of the media today derives from and reveals a similar abdication. For too many years, congressmen gave the president too much room, asked too few questions, took too few initiatives of their own. Many of them were content to be courtiers, merely resentful when they were excluded from the court, with little sense of an alternative politics. No doubt, they have enjoyed Nixon's fall, but they are also frightened by it and unable to respond in any coherent way.

The brief career of Sam Ervin as a republican hero suggests the possibilities of congressional action and also the weakness of Congress as an institution. Even the effectiveness of the Ervin Committee derived in large part from its adoption by the media: the senators and their witnesses put on a good show. And when they ceased to do that, the Committee ceased at the same time to command attention—almost as if it had gone underground. No doubt, it will emerge one day with legislative proposals of some sort, useful proposals, presumably, but also moderate and minor. There seems no reason to expect decisive action from the Congress. And yet, within our constitutional system, there is no other place from which decisive action can come. The media can only amaze us, stir us up, prepare us for action, but they can't speak for us, and they can't act in our name. In their hands, politics has become a spectator sport; Watergate is merely a scandal, and we are fascinated and delighted by every new detail in its exposure. But what is to become of the public business?

Perhaps the greatest problem is this: there is no opposition party in the United States today. That is why we have no way of responding to Watergate and no organized pattern of political work in which we can involve ourselves. Never has American politics seemed so anarchic as it does right now when individual candidates and would-be candidates scurry about trying to distance themselves from the scandals and establish their personal cleanliness. They cannot hope for more than marginal success—one consequence of Watergate as a public spectacle is a spreading cynicism about all politicians—but marginal success will probably be enough to win the next election.

The only way to combat the general cynicism, however, is for some group of men and women to offer an explanation of the present crisis and point the way to an alternative politics. The logical place for that to happen is the Democratic party, but where is the Democratic party? Who can find it? No doubt, there is a great reluctance among its possible leaders to seem to be seeking partisan advantage in a time of troubles. The result of that reluctance is that only personal advantages are sought. So everyone connives in fostering the proposition that the wickedness of a few men, and nothing more, lies at the root of the crisis, and that the personal decency of a few men, and nothing more, will suffice to deal with it. Face-to-face with Claudius, we are waiting for the Prince.

I don't want to deny the value of throwing the rascals out. That is a good thing to do, not only because of the pleasure it would involve. And there is something especially impressive about the machinery provided by the Constitution for throwing out the president himself. In a short time, we will know whether or not there is to be an impeachment. Right now, the odds seem heavily against it. It would require a kind of cohesiveness and collective resolution which neither Congress as a whole nor the Democratic party in Congress has yet demonstrated. But how splendid it would be to see those rusty constitutional gears grinding into motion! How valuable the formality and solemnity of the law would be at this moment, after years of presidential lawlessness! And yet impeachment is only a ritual; one wants most what it would symbolize—the reappearance of a republican politics. That means, of presidential plainness, congressional energy, popular participation. For the moment, there is little sign of any of these. Political virtue is passively represented by those ordinary citizens who —it takes no courage—tell George Gallup week after week that they don't approve. Perhaps they are also not resigned. But what alternatives do they have?

(1974)

4

Social Origins of
Conservative Politics

I

"Now begins the fight for the soul of the Republican party"—
so a television commentator announced over and over again
(as television commentators will do) on election night. It was
not quite an accurate statement. American political parties
don't have souls; they are very lowly organisms, loose associations of their parts, which work closely together only during
presidential campaigns. What happened in 1964 was that even
this association broke up. Faced with the disaster of the Goldwater candidacy, the party went through a process of protective disintegration and turned itself into a collection of
autonomous state and local bodies. It was replaced as a national
entity by the Goldwater movement, whose leaders boasted
of their self-sufficiency and made no effort at all to renew the
associational ties. Now that the movement has been so over-
whelmingly defeated, the party will have to be reconstituted
through the usual process of bargaining and compromise
among its various parts.

The defeat of the Goldwater movement is without precedent in American political history. But so is the Goldwater movement without precedent, and it is by no means clear that the defeat was so utterly crushing as to destroy the movement. Its candidate, after all, polled 25 million votes, almost 40 percent of the total, rather more than the opinion polls suggested he would get. Many Republican party workers and militants remain loyal to a right-wing politics and are probably willing to commit themselves to a long period of "principled" opposition. If the movement can no longer hope to replace the party, it also seems unlikely that the party can dispense with the movement. The Goldwaterites will take their place as one among its constituent elements, but they will be a novel element for an American party to contain: not a local organization, founded on place and patronage, but a disciplined, ideological faction.

In retrospect, it is obvious that they ought to have settled for factional status before the election. Though it looked to be growing, the Goldwater movement simply could not sustain a bid for exclusive national power. The strains and tension of American life, racial conflict and cold war frustration—all of which seemed to be driving voters toward the Right—were not sufficiently acute to produce anything like an electoral majority. Goldwater's candidacy came, it seems to me, too soon. Right-wing politics waits on crisis; the candidate did not wait. The best example of the amateur enthusiasm and ineptitude of the 1964 Republican campaign was the decision to run that campaign in 1964.

But an ideological movement doesn't require victories in quite the same way as does a conventional American party. It depends for survival and growth on the openness of social groups to its ideology, on the availability of men and women whose fears it can exploit or whose claims it can assert and legitimate. Today there are a great many Americans, but not enough, available for a right-wing politics: will there be more or fewer of them in 1968, 1972, 1976? The future of the Goldwaterites doesn't depend on their success or failure in the intra-party struggle of the next few months or years (in which they are likely to fare badly), but rather on the shape of American society in the next few decades. And because the "great society" that President Johnson holds out as our future has as yet no shape at all, it remains an open question whether the movement can survive, slowly build its strength, and even bid for power once again if racial conflict or cold war tensions reach a crisis point.

II

What is it that makes Americans join the radical right? During the past year we have been treated to a great deal of sociological speculation, both academic and popular, aimed at answering that question. In what has become the typical American fashion, however, the range of speculation has not been very wide; nor has there been much disagreement. Discussion has centered around a domesticated version of the theory of traditionalist revolt, first proposed by Talcott Parsons in a famous essay on Nazism.[1] Whenever Goldwaterism is traced to status frustration or resentment, or described as a "politics of nostalgia" or a revulsion from modern complexity, or linked to fundamentalist Protestantism, or explained simply as an appeal to the old-fashioned American virtues, it is Parsons's theory, or some revised or popularized version of it, that is being invoked. Transferred to an American setting, the theory loses some of its sting: traditions rooted in the soil of Ohio are far less ominous than traditions rooted in the soil of Prussia. And that is probably the reason for its popularity. It makes possible just the right amount of self-reproach and self-congratulation. It's a terrible thing that so many Americans should be dissatisfied with the modern world and that a man like Goldwater should come anywhere near the presidency.

[1] "Democracy and Social Structure in Pre-Nazi Germany," in *Essays in Sociological Theory* (Glencoe, Ill., 1954), pp. 104–123.

But it's a wonderful thing, and a tribute to the strength and stability of American modernity, that he should be so soundly defeated.

This theory of Goldwaterism has been argued in some detail in recent issues of *Encounter*, first by Richard Hofstadter, the Columbia University historian, and then by S. M. Lipset, the Berkeley sociologist.[2] Hofstadter sums it up:

The Goldwater movement is a revolt against the whole modern condition as the old-fashioned American sees it—against the world of organisation and bureaucracy, the welfare state, our urban disorders, secularism, the decline of American entrepreneurial bravura, the apparent disappearance of individualism and individuality, and the emergence of unwelcome international burdens. Although its enthusiasts like to think of themselves as conservatives, their basic feeling is a hatred of what America has become. . . .

Lipset describes the revolt of old-fashioned Americans as a case of WASP "backlash." This backlash is directed not only or most significantly against the Negro, but, again, against the cosmopolitanism and modernity of our urban culture and our national politics. Goldwater appears as the last embodiment of American Babbittry, the last spokesman for the Puritan ethic and the frontier spirit—not, of course, in their original forms but as these have been incorporated into the life and mythology of the small town. Hence his hatred of big government and the welfare state and his endless preachments about self-reliance and masculine determination. Hence also, since Babbitt is so hard pressed these days, the paranoia and resentment that lie just beneath the rugged surface of right-wing politics.

All this implies a very complacent view of the shape of

[2] Richard Hofstadter, "Goldwater and His Party," *Encounter* 37, no. 10 (October 1964); Seymour Martin Lipset, "Beyond the Backlash," *Encounter* 37, no. 11 (November 1964).

American society and an extraordinarily optimistic prediction about the future of Goldwaterism. It is worth examining carefully. The theorists of traditionalist revolt see an America divided into two parts, one visibly in decline, the other happily ascendant. The first is a WASP world of independent farmers and small businessmen, realtors and insurance salesmen, doctors and lawyers. Its inhabitants are loyal churchgoers, these days chiefly Methodist and Baptist, though the upper reaches of small town society will still be Presbyterian or Episcopalian. Puritanical in their ways or at any rate in their speech habits, these are the people who answered *no* to all Dr. Kinsey's questions and who presumably responded eagerly to Goldwater's vague gestures against "moral decay." They are economically independent and accustomed both to high status and considerable power in American society. For many years, they have dominated our state legislatures and even the Congress.

WASP interests and ideals, and more recently WASP fears, have played a considerable part in national politics. The antimonopoly legislation of the early twentieth century was a victory for small town Americans: they have always opposed to the impersonal power of Eastern finance, as now to the impersonal power of Eastern government, an ideology of private enterprise and individual rectitude. Prohibition was their achievement, their great effort to turn America into a Puritan society. The closing down of immigration in the 1920s, during the presidency of Warren Harding of Marion, Ohio, was their achievement, a desperate attempt to maintain their supremacy in American life (Lipset describes it as the most recent example of WASP backlash).

Since the 1920s, however, this first America has been in radical decline. The ruin of so many farmers and businessmen during the depression, the drift of millions to the cities, the consolidation of the immigrant groups into powerful electoral blocs, the rise of the industrial unions, the growth of the national government, and the shift of power within the government from Congress, which small town America can still control, to the presidency, which it can't—these have been devastating blows to its power and prestige. At first these blows were resisted with a genuinely conservative politics, best represented during the 1940s and early 1950s by Senator Robert Taft. But by now, so the theory goes, the balance has shifted so sharply against the WASPs that Taft conservatism has been replaced by a troubled nostalgia and a militant reaction, the two together represented by Goldwater.

But the strength of the Goldwater movement does not rest in the small towns—where Nixon's more conventional Republicanism had far greater appeal. It is among small town Americans who have immigrated to the big city that the encounter with modernity is most sharp and the traditionalist revolt most powerful. Ever since the 1920s our cities have been swollen with immigrants not from Europe, but from our own Midwest and South. At first, these people went into the factories, were organized and educated by the unions, and so brought into some comprehensible connection with the modern world, the second America. But more recently, they have drifted into the lower levels of corporate bureaucracies or into the rapidly expanding service industries, still largely unorganized. There they are not absorbed into or educated by urban communities, but simply *included* in the new middle class. They discover urban culture and society as a hostile environment. They cannot "adjust" to the tempo of city life, to bureaucratic impersonality, to the easy familiarity that replaces friendship. They are repelled by the sophistication and secularism of old city dwellers, by the wantonness of urban women and the aggressiveness of the lower classes.

Right-wing politics is a response to the impersonal and competitive world of the modern city. "Americanism" recalls the innocent patriotism and the moving ceremonial routines of small town life and bolsters a collapsing self-esteem. The ideology of free enterprise reinforces an otherwise failing sense of personal energy and power. Puritanism expresses the fear and dislike of the corrupting city, America's Sodom and

Gomorrah. Anti-unionism, contempt for the poor, the unemployed, the blacks, represent a desperate self-assertion, a claim to an old supremacy. And finally, paranoia, the fear of plots and treason, is simply a way, for many perhaps the only way, of explaining the decline of the old America.

According to the theorists of traditionalist revolt, then, the Goldwater movement mobilizes two groups: the small town Americans in the small towns; the small town Americans in the big cities. It directs this alliance against the modern world—liberal, pluralist, cosmopolitan—which includes in some strange and happy combination all the enemies of WASP America: governmental bureaucracies, interlocked with Eastern big business through complex processes of nonregulation; big business, connected in turn with the unions through stable patterns of collective bargaining; unions and organized minority groups, whose memberships overlap, both linked in turn to governmental bureaucracies through patronage and welfare politics, etc., etc. But it is hardly necessary to analyze this second America, which is experienced, we are told, simply as a vague and conglomerate threat and blamed indiscriminately for anything that goes wrong.

III

Now the crucial bias of liberal sociologists and journalists is just the opposite: that modernity and all it includes is good, and that modern America is indeed capable, as President Johnson promises, of incorporating every American and satisfying every one of us. And so they see Goldwaterism as a phenomenon of a transitional period in our history. Their view is optimistic because it places the source of right-wing strength in declining social groups or among recently mobile men and women who only need a little time to adjust themselves to their new positions in urban society and to learn the sophisticated styles of urban culture. The theory of traditionalist revolt only barely conceals a self-confident affirmation of historical progress. Thus Lipset: "Finally, it cannot be stressed too strongly that the politics of the backlash, whether religious-cultural, economic, or racially motivated, is a consequence of the fact that the United States is becoming more liberal. The groups now reacting with such desperation are desperate precisely because they are growing less influential and less numerous."

The election results don't make it easy to pass judgment on the notion of a WASP backlash. Goldwater seems to have done less well than Nixon did four years ago among every ethnic group and social class in every part of the nation (with the exception of the deep South). The results are so uniform as to be sociologically impenetrable—at least until political scientists publish careful and detailed comparisons of the swings in particular precincts. And among the 25 million people who voted for Goldwater it is not at present possible to distinguish habitual Republican voters and reluctant party loyalists from committed supporters of the right-wing movement. We still don't know, in other words, just who the Goldwaterites are. Nevertheless, there seem to me to be some very good reasons for arguing that the theory of traditionalist revolt provides at best only a partial explanation of the Goldwater movement and does not warrant the optimism which has been based upon it.

Begin first with the party convention in July. Several reporters wrote at the time that they had gone to San Francisco expecting to find the Republican party in the hands of frightened and fanatical small businessmen, old and pensioned-off WASP ladies, old and pensioned-off air force colonels, used car salesmen from Ohio and Indiana, etc. What they found in fact was quite different: suave and aggressive college graduates, young corporate executives on their way up, upper middle-class professional men and women, confident, tough-minded, power-hungry. The other kind was there too, of course, hysterical in the galleries; but the core of the movement seemed to be made up of people who, whatever their problems, can only be described as modern Americans. No doubt their adherence to the Goldwater movement can be accounted for by one or another sort of social pathology, but I want to argue for a moment that it can best be explained not as a politics of resentment or hostility but as a *politics of interest*. It is the politics of a new upper middle class and a new upper class in the United States whose status rests at present only on its money and is threatened or thought to be threatened by high taxation and the very moderate egalitarian tendencies of the welfare state. Writing in *Dissent*, Irving Howe has summed up this view of Goldwaterism:[3]

Muddled nostalgia, homestead economics, day-dream brinkmanship all enter the Goldwater complex. But there is also a hard ideology, a precise focus of interest, and no one should have any trouble in identifying these: they animate a politics of reaction. This politics rests in part on the social selfishness of large segments among the upper middle class which consciously disdain the sentiments of "welfarism" and are prepared to let the poor and the ethnic minorities stew in silence or, if noisy, be slugged into submission. This politics rests on the desire of powerful men—Ralph Cordiner is not a mid-western automobile salesman, George Humphrey not a malaise-smitten petty bourgeois, H. L. Hunt not a bewildered storekeeper—to call a halt to social reform and then slowly to push it back.

Goldwaterism suits the needs of a group of Americans that may well have been growing and has certainly become more noisy over the last few years—though it is still a minority group and a fairly distant noise. The success of capitalism in the United States has tended to increase the number of people who have no interest whatsoever in welfare policies and

[3] "The Goldwater Movement," *Dissent* 11, no. 4 (Autumn 1964).

can see them only as a drain on their own income. Set free from every kind of communal support, they regard the very idea of communal responsibility as an attack upon their cherished independence. Goldwater's speeches about self-reliance don't merely reflect the old myths of free enterprise, but also a very new sense—felt even among American teen-agers—of *having money enough and of one's own*. And from this there seems to follow an utter inability to understand the troubles of people who don't have money enough.

The welfare state, after all, is not a purely modern phenom-enon. It rests upon strong traditions of communal coopera-tion and support, traditions best preserved during the past hundred years in the unions and among ethnic and religious minority groups. Insofar as individual prosperity is achieved at the expense of these communities, insofar as it involves the breakdown of old patterns of economic and social cooperation, it produces Goldwaterites, that is, men and women whose only politics is selfishness and self-aggrandizement. They are cer-tain that they have "made it" and what is more, that they have "made it on their own." That is undoubtedly the classical form of entrepreneurial self-esteem, but it now extends beyond the world of commercial or industrial entrepreneurs. It is a belief cherished throughout the *modern* social class which C. Wright Mills described in *White Collar*, whose members are, so to speak, entrepreneurs of the self. They have taken risks, broken away from their families, rushed off to new cities like Los Angeles and Dallas, sold their personalities and their talents when they had nothing else, lived on time and credit, moved again and again, accepting rootlessness and loneliness as the price of economic advance. And now, when so many of them have won whatever rewards the system offers, they are, not un-naturally, most unwilling to share those rewards. No one shared with them. Whatever hidden support they received from the welfare state or from the war or the war economy was, after all, well hidden. It never represented any concrete relief from the oppressions of the rat race.

The Goldwaterites are the possessors of recent wealth. New men with new money, the liberal sociologists argue, are always insecure and possessive; but they will grow old and learn the service ideology of upper class conservatism, and their money will mature until, like Rockefeller money, it is available only for good causes and liberal politics. What is strange here is the notion that all wealth must grow old in the same way. It is of course true of all right-wing movements that they begin without the support of established financial and industrial interests, for these have made arrangements with the government and the unions which they are unwilling to jeopardize for the sake of such an uncertain payoff as groups like the Goldwaterites can offer. But it is not inevitable that the newer wealth will, with time, settle into the old arrangements. Right-wingers today obviously are attempting to alter those arrangements radically in their own favor. Election night analyses suggest tentatively that Goldwater did best among the wealthy and in some of the well-to-do suburbs of our larger cities: with good reason. The size of his defeat will undoubtedly force many of his supporters to seek some accommodation with the established interests, but it will not mark the end of their self-aggrandizement nor, necessarily, of their political involvement. Lipset's comparison of the Goldwater movement to French Poujadism seems to me especially inapt.

IV

There is something more to be said, however, for the interests even of Goldwaterites extend beyond economic selfishness. Public opinion polls indicate that during the campaign Goldwater won his greatest response when he talked of "moral decay." Now this extraordinary concern with morality among his followers was not, any more than their demand for self-reliance, a mere expression of old but fading Protestant virtues.

Goldwater's emphasis in the last weeks before election day on pornography and sex was undoubtedly a bid to touch off the ugly puritanical prurience which runs so deep in the American character, and the outcry which he sought to raise about crime in the streets was, of course, a bid for white backlash support. But it was something more: it was an expression of tensions and strains in the lives of his own supporters—and of many other people who, for one or another reason, chose not to support him.

In his *Encounter* article, Lipset cites an interesting statistic: a disproportionate number of newcomers to the city of Los Angeles figure among local supporters of the John Birch society. He interprets this as a demonstration of his thesis that small town Protestant immigrants to the cities express their uneasiness and maladjustment through right-wing politics. But it may be that it is not the initiation to the big city, not the difficulties of urban life, that produce this reaction, but rather the quality of the city to which one is being initiated and the particular sorts of demands that a particular sort of urban life makes upon individuals. It may be that extreme strain, mobility, competition, anxiety, and disorder are features not merely of a transitional period on the way to modernity, but of modernity itself —or at any rate of one of the several alternative modernities currently being shaped in the United States: a particularly raw, because very advanced, form of capitalism, unmitigated by the solidary structures of an older urbanism. If this is so, then Goldwaterites even while defending their new and prosperous way of life are also the first victims of the way they live. They prefer to talk of crime in New York or Washington, but cities like Phoenix and Dallas, governed for some years now by Goldwater types, apparently have higher crime rates.

Moral decay is a complex business. It is very much bound up with the shattering of the old communities and institutions which uphold standards. And it is usually not the people who still live in the old communities and still believe in the old standards who are most upset by the spectacle of collapse, but rather the people who find themselves among the ruins and become the hapless defenders of communal ruination. Goldwaterism is itself an agent of the moral decay it condemns. But that decay is real, manifest in urban restlessness, privatism, and mutual indifference, and it is certainly not obvious that this sort of thing is being left behind as America becomes more modern. Instead it seems, as the Goldwaterites say, to be growing. However delusory right-wing solutions may be, Goldwaterism could well grow with it, unless there is some major effort to find alternative solutions and to recreate some sort of urban communal life.

Modern selfishness and modern discontent are far more significant in Goldwaterism than traditionalist revolt. From private affluence there have emerged new claims for private privilege and power. And because private affluence produces no new communities within which such claims can be adjudicated, there have also emerged new patterns of anxiety and moral disorder. All this is especially dangerous today, when other claims are being made upon our society and other disorders being bred. The demands of the blacks do not bring them into conflict with traditionalist WASPS, but precisely with those modern entrepreneurs of the self who have "made it" since the war. And this conflict is likely to be intense and deadly because modern Americans are incapable of understanding the troubles that blacks experience *collectively*, the group disadvantages which can only be remedied by collective action.

It was the anticipation of white and not WASP backlash that probably determined the mistiming of the Goldwater campaign. This backlash did not materialize, partly for reasons that do credit to the American voter, but chiefly because the civil rights movement has not yet made sufficient progress to threaten anyone. But it is almost certain that in the immediate future the civil rights struggle will offer great opportunities to right-wing politicians. They do not have to defend segregation, but only proclaim the absolute egotism and the freedom from all communal control that were Goldwater's campaign themes and that would make integration impossible. The defeat of fair housing legislation in the California referendum suggests their strategy: the constitutional amendment which the voters approved stated simply that citizens might sell their homes without restraint or restriction. Here was a triumph for Goldwaterism, even though the candidate himself went down to ignominious defeat. There will be other such triumphs, though how many it is impossible to say, for the precise strength of Goldwater's America has not yet been measured. It will only be known when the hopes for privilege and power that private prosperity has bred are decisively challenged.

(1964)

5

Nervous Liberals

I

A genuine conservatism expresses a sense of crisis and imminent or actual loss. Its tone is perfectly caught in the opening lines of Richard Hooker's *Laws of Ecclesiastical Polity*, where Hooker explains his purpose in writing: "Though for no other cause, yet for this, that posterity may know that we have not loosely through silence permitted things to pass away as in a dream. . . ."[1] And, more stridently, in the gothic prose of Edmund Burke: "But the age of chivalry is gone. That of sophisters, economists, and calculators, has succeeded; and the glory of Europe is extinguished forever. Never, never more shall we behold . . ." etc.[2]

Our own neoconservatives express a neo-sense of crisis and loss. Though they sometimes write in the gothic mode, they cannot approach Burke's wholeheartedness. For they themselves stand in the ranks of the economists and calculators.

A review of Peter Steinfels, *The Neo–Conservatives: The Men Who Are Changing America's Politics* (Simon and Schuster: New York, 1979).

[1] *Of the Laws of Ecclesiastical Polity*, preface.
[2] *Reflections on the Revolution in France*, ed. Conor Cruise O'Brien (Harmondsworth, England, 1968), p. 170.

They are committed to the arrangements and processes that cause the transformations they bewail. As Peter Steinfels writes in his excellent study of neoconservative thought, they live with a "basic dilemma": "The institutions they wish to conserve are to no small extent the institutions that have made the task of conservation so necessary and so difficult."

What is the nature of the "crisis" that American neoconservatives have been complaining about? Among the writers Steinfels considers, the crisis is differently described and with very different degrees of analytical rigor. I can only suggest a rough and quick summary. Steinfels provides a careful analysis, skeptical, but always true, I think, to the best of their arguments. The crisis is first of all a collapse of authority in governments, armies, universities, corporations, and churches. Old patterns of trust and deference have broken down. Political leaders, military officers, factory foremen cannot command obedience; professors cannot command respect. Alongside this is a radical loosening of social bonds in communities, neighborhoods, and families—perhaps best summed up in the common metaphor of "splitting." Once only Protestant sects and radical political movements split. Now families split, couples split, individuals split. Splitting is the ordinary and casual way of breaking up and taking one's leave, and leave taking is one of the more remarkable freedoms of contemporary society.

Finally, there is a deep erosion of traditional values, not only deference and respect, but moderation, restraint, civility, work. All this makes for a pervasive sense of disintegration. It creates a world—so we are told—of liberal decadence, of rootless, mobile, ambitious men and women, free (mostly) from legal and social constraint, free too from any kind of stable intimacy, pursuing happiness, demanding instant satisfaction: a world of graceless hedonists.

This picture obviously depends upon implicit comparisons with some older and different social order and, as Steinfels makes clear, the precise historical reference points are rarely given. So the picture is crudely drawn, a disturbing combination of insight and hysteria. As expressed in the writings of Irving Kristol, Robert Nisbet, Aaron Wildavsky, Samuel Huntington, Daniel Moynihan, S. M. Lipset, Nathan Glazer, and Daniel Bell—professors or former professors all—it has to my mind an initial implausibility. It relies too heavily on the experience of the late sixties and hardly at all on the experience of the late seventies. The authority of presidents, in the aftermath of Vietnam, Cambodia, and Watergate, may still be precarious, and understandably so, but the authority of professors seems fully restored. That probably has more to do with the economy than with our own virtue or pedagogical success. Still, students have never in the past twenty years been as deferential as they are today. Conservatives are supposed to dwell happily in the past; the present is torment for them. Our neoconservatives dwell miserably in the past, reliving every undergraduate outrage; the present might be a relief.

But let us accept their vision, or at least take it seriously. That is Steinfels's strategy, and it is surely right. These neoconservatives are eminent scholars and intellectuals; they are widely read (because they have interesting things to say); they have ready access to foundations and government agencies. Though they differ among themselves in ways I mostly won't be able to explain in this review, they constitute a common and increasingly influential current of opinion. Steinfels claims that they have created at last that "serious and intelligent conservatism that America has lacked, and whose absence has been roundly lamented by the American Left." Though the adjectives are right, the claim is dubious, for these writers, on Steinfels's own reading, have not resolved the basic dilemma of conservative thought; nor are they genuinely committed to the world that is passing away. Still, their argument is worth pursuing. Even if we don't experience the contemporary crisis with the intensity conveyed in their essays and books, we do after all have intimations of its reality.

It is odd, however, to represent that reality as the decline of liberal civilization. I would suggest instead that what we are living with today is the crisis of liberal triumph. Capitalism, the free market, governmental *laissez faire* in religion and culture, the pursuit of happiness: all these make powerfully for hedonism and social disintegration. Or, in different words, they open the way for individual men and women to seek satisfaction wherever they can find it; they clear away the ancient barriers of political repression, economic scarcity, and social deference. But the effects of all this are revealed only gradually over decades, even centuries. Today, we are beginning to sense their full significance.

"The foundation of any liberal society," Bell has written, "is the willingness of all groups to compromise private ends for the public interest."[3] Surely that is wrong; at least, it is not what leading liberal theorists have told us. The root conviction of liberal thought is that the uninhibited pursuit of private ends (subject only to minimal legal controls) will produce the greatest good of the greatest number, and hence that every restraint on that pursuit is presumptively wrong. Individuals and groups compromise with one another, striking bargains, trying to increase or "maximize" private interest. But they don't compromise for the sake of the public interest, because the public interest—until it was resurrected as *The Public Interest*—was not thought to be anything more than the sum of private interests. From this maximizing game, however, large numbers of men and women, the majority of men and virtually all women, were once excluded. They were too poor, too weak, too frightened. It is this exclusion, I suspect, that figures in neoconservative writing as the moderation and civility of times gone by. And what is called hedonism is in reality the end of that exclusion as a result, largely, of economic expansion, mass affluence, and a "liberating" politics that does little more than exploit the deepest meanings of *laissez faire*.

Hedonism certainly isn't new. One has only to think of

[3] "The Public Household," in *The Public Interest*, no. 37 (Fall 1974), p. 46.

America in the gilded age or in the 1920s. Nor is it newly cut loose, as neoconservative writers frequently suggest, from the Protestant ethic. If one wants to understand the consumption habits of earlier Americans, one would probably do better to read Veblen than Max Weber. But it is true, and important, that hedonism as a way of life is newly available outside the upper classes. More people pursue happiness, and they pursue it more aggressively, than ever before. Workers, blacks, women, homosexuals: everyone is running. Everyone's entitled. It makes for a lot of jostling, but isn't this the fulfillment of the liberal dream? No one reading Hobbes and Locke, and foreseeing the economic expansion of the years since they wrote, would be surprised. And yet how much we miss those old social gospel Christians, populist reformers, socialist agitators, who forgot themselves and pursued other people's happiness! And how much our neoconservative colleagues miss all those men and women who never realized that they had a right to run!

What is true in the economy is also true in politics. "The effective operation of a democratic political system," writes Samuel Huntington, ". . . requires some measure of apathy and non-involvement on the part of some individuals and groups. In the past, every democracy has had a marginal population, of greater or lesser size, which has not participated in politics." This marginality, "inherently undemocratic," is nonetheless one of the conditions of democratic success—or at least of governmental effectiveness.[4] The argument might be put more baldly. In the past, government was able to respond effectively to the demands of the powerful and the well-organized, but it is threatened (and authority and civilization with it) when demand is universalized, when everyone gets into the political act. Yet liberal democracy tends toward universality of exactly that sort. What is to be done?

A similar story can be told about religious life. *Laissez faire* in religion works wonderfully when it is a matter of creating a

---

[4] "The Democratic Distemper," in Nathan Glazer and Irving Kristol, eds., *The American Commonwealth: 1976* (New York, 1976), p. 37.

structure within which well-established creeds, with well-disciplined adherents, coexist. But as the established religions slowly fade away (in an atmosphere of radical disestablishment, hostile to institutional pretensions), they are replaced by a proliferation of sects and cults, and the stability of the general structure is strained. All sorts of people want to be saved, right now, and as there are many paths to the house of the Lord, so there are many hawkers selling maps. Contemporary sectarianism is simply the latest product of the market. Its leaders combine charisma and hustle, and one can read in their activities all the signs of entrepreneurial energy and, sometimes at least, of consumer satisfaction. Watching the Hare Krishna people on the streets of New York or Cambridge, I probably have feelings very similar to those of a seventeenth-century Puritan minister (the neoconservatives probably feel like Anglicans) listening to a Ranter or a Fifth-Monarchy man. But I still value religious freedom—as do the neoconservatives. And so again: what is to be done?

II

In an impressive sentence, Irving Kristol has written that bourgeois society lived for years off "the accumulated capital of traditional religion and traditional moral philosophy"—capital it did not, as Steinfels emphasizes, effectively renew. The point can be generalized. Liberalism more largely, for all its achievements, or as a kind of necessary constraint on those achievements, has been parasitic not only on older values but also and more importantly on older institutions and communities. And these latter it has progressively undermined. For liberalism is above all a doctrine of liberation. It sets individuals loose from religious and ethnic communities, from guilds, parishes, neighborhoods. It abolishes all sorts of controls and agencies of control: ecclesiastical courts, cultural censorship, sumptuary laws, restraints on mobility, group pressure, family bonds. It creates free men and women, tied together only by their contracts—and ruled, when contracts fail, by a distant and powerful state. It generates a radical individualism and then a radical competition among self-seeking individuals.

What made liberalism endurable for all these years was the fact that the individualism it generated was always imperfect, tempered by older restraints and loyalties, by stable patterns of local, ethnic, religious, or class relationships. An untempered liberalism would be unendurable. That is the crisis the neoconservatives evoke: the triumph of liberalism over its historical restraints. And that is a triumph they both endorse and lament. A small illustration: Kristol writes angrily that in the contemporary world, "to see something on television is to feel entitled to it." "He nowise hints," Steinfels comments, "that this is exactly the reaction that someone has intended, in fact spent considerable sums of money to create." Free men and women, without strong roots in indigenous cultures, are open to that sort of "creativity," and liberalism by itself offers no protection against it. Do the neoconservatives propose to protect us? Though Kristol has urged the censorship of pornography—one more product of the free market—he has not, so far as I know, urged the censorship of advertising. Still, he is uneasy with the consequences of freedom.

Neoconservatives are nervous liberals, and what they are nervous about is liberalism. They despair of liberation, but they are liberals still, with whatever longing for older values. They remind me of a sentence about Machiavelli hastily scrawled in an undergraduate's blue book: "Machiavelli stood with one foot in the Middle Ages, while with the other he saluted the rising star of the Renaissance." That is the way I think of Irving Kristol. He stands with one foot firmly planted in the market, while with the other he salutes the fading values of an organic society. It is an awkward position.

It is also, intellectually and politically, a puzzling position. In recent years, the main tendency of neoconservative writing has been a critique of state intervention in the economy and of expanded welfare programs of the Great Society sort. In magazine articles, foundation studies, and *Wall Street Journal* editorials, we are repeatedly shown public officials struggling to respond to the cacophony of demand generated by mass democracy, struggling to do (badly) for men and women what they once did (better) for themselves and one another. Like Prince Kropotkin, the neoconservatives dislike the state (unlike the Prince, not the police) and they believe in mutual aid. They value those old communities—ethnic groups, churches, neighborhoods, and families—within which mutual aid once worked. Or supposedly worked: once again, I don't know the historical reference of the argument. In any case, the basic dilemma remains. For they are committed at the same time to the market economy whose deepest trends undercut community and make state intervention necessary. To put the argument most simply: the market requires labor mobility, while mutual aid depends upon local rootedness. The more people move about, the more they live among strangers, the more they depend upon officials.

Today, that dependency is genuine and pervasive. Capitalism forces men and women to fight for the welfare state. It generates what is indeed a very high level of demand for protection against market vicissitudes and against entrepreneurial risk taking and for services once provided locally or not at all. It is a common argument among neoconservatives that this demand "overloads" the welfare system, which cannot provide the protection and services people have come to expect. Trapped by the necessities of electoral and pressure group politics, political leaders promise more and more social goods. In office, inevitably, they fail to deliver; popular respect for government declines; the crisis deepens. Perhaps this view expresses some ultimate truth about the welfare system. With Steinfels, I am inclined to doubt that it expresses any immediate truth. It justifies, as he says, a politics that holds too quickly and without sufficient reason that minimal decency in, say, health and housing is simply beyond the reach of our (discredited) officials.

But it is not a part of Steinfels's project to pursue such disagreements in detail. He is concerned with exposition and analysis. Suppose, then, that the overload argument is right. The long-term effect of liberalism (and capitalism and democracy) is that too many people want too much too quickly. What follows? It isn't possible to drive individuals and groups back into an older condition of passivity, deference, and marginality. I sometimes detect a hankering after the days of the "respectable poor" among the neoconservatives, but the repression that would be necessary to bring those days back is not a part of their programs. These are liberals still, however nervous. Indeed, it is not clear that there is a coherent program either for interdicting overload or for coping with it.

At this point, articles in *The Public Interest*, a journal whose editors boast of their hardheadedness, turn preachy. "Less marginality on the part of some groups," writes Huntington, "needs to be replaced by more self-restraint on the part of all groups."[5] Yes, indeed. But what is going to persuade all those individual and collective selves to set limits on their demands? What sets of beliefs, what political movements, operating within what sorts of institutional structures? Unless answers are provided for questions like these, answers that give some bite to the crucial phrase in Huntington's sentence—"on the part of *all* groups"—neoconservatism is likely to collapse, as Steinfels writes, into "the legitimating and lubricating ideology of an oligarchic America . . . where great inequalities are rationalized by straitened circumstances. . . ."

Among neoconservative writers, Daniel Bell comes closest to dealing with these questions—though he deals with them in a way that raises doubts about his own conservatism. In fact, he has kept his intellectual distance; it is Steinfels who makes the connection, arguing for the primacy within the corpus of

[5] "Democratic Distemper," p. 37.

Bell's work of his attack on modern culture and mass hedonism. Certainly, Bell is as worried as his friends on *The Public Interest* are about the loss of *civitas*, "that spontaneous willingness . . . to forgo the temptations of private enrichment at the expense of the public weal," and he is as loathe as they are to tell us when it was that such temptations were actually forgone.

Almost alone among neoconservatives, however, Bell is prepared to recognize that *civitas* depends upon a pervasive sense of equity and that equity in America today requires greater equality and a more effective welfare state. When Bell calls himself "a socialist in economics," he is marking a difference between his own work and that of his friends that is worth stressing. Steinfels points out that Bell's socialism is rarely reflected in his writing on economic institutions; it is programmatically thin; and it is accompanied by reiterated expressions of hostility toward egalitarian radicals. But his argument for "the public household" does at least suggest some way of reincorporating liberal values in new communal structures. The alternative is to make a politics out of nervousness itself, a crackling, defensive, angry, unfocused politics—as much of neoconservatism is.

III

Equality is a specter that haunts the neoconservative mind, and Steinfels writes about the haunting with great insight in what is probably the strongest chapter of his book. Like him, I have some difficulty identifying the object which the specter is supposed to represent. Is it the New Left, long gone, or the civil rights movement, or the black and feminist campaigns for affirmative action? All these taken together have hardly carried us very far (any distance at all?) toward that "equality of outcomes" which Nisbet, Kristol, Glazer, Bell too, regard as the clear and present danger of contemporary political life. These writers put themselves forward as defenders of meritocracy (Bell, characteristically, of a "just meritocracy," within which those on top cannot "convert their authority positions into large, discrepant, material and social advantages over others," a qualification for which he should probably be denounced in *Commentary*). But if their goal is "a career open to merit," then surely they must sense that real progress has been made in that direction in the past several decades, and not through their efforts. The advance has largely been forced by the egalitarians they attack. And most of them, the mainstream of blacks and women certainly, would be more than happy with a genuine meritocracy.

But would the neoconservatives be happy with that? Who are the meritocrats anyway but rootless, ambitious men and women, cut loose from traditional communities, upwardly bound, focused on the state? And isn't it these people, unsure of their present position and their final destination, full of status anxieties, envious of older elites and established wealth, who—according to neoconservative polemics—carry in their hearts and minds the germs of a radical egalitarianism? Here again is the neoconservative dilemma. As these writers yearn for lost communities, so they yearn for lost hierarchies and stable establishments. How else can authority regain its luster except by being embodied in a class of men (and women too, if necessary) confident of their place, trained for power and public service, secure against competitors? But meritocracy undermines all such classes. Whether it is happiness that is being pursued, or position and office, the scramble that results leaves no one secure or confident. All the neoconservatives are meritocrats in practice as well as in theory. They have earned their places in academic and political life. But they are uneasy with their fellows. This uneasiness is expressed in the virtually incoherent doctrine of the "new class."

Steinfels devotes three chapters to his strange argument that figures so largely in neoconservative (and also in neo-Marxist) thought. The subject is important because it is in writing about the "new class" that neoconservatives give us the clearest sense of who they think they are and who they think their enemies are. Unfortunately for social analysis, both they and their enemies seem to belong to the "new class"—which is therefore described, alternately, with warm affection and deep hostility. The political universe of neoconservatism is narrow: it consists of students, professionals, technocrats, bureaucrats, and intellectuals. The old bourgeoisie is gone, along with liberal civilization; the workers are summoned up only when it is necessary to remind them of the importance of wage restraint. Politics, as Steinfels writes, is a "war for the new class." He might have added, it is a civil war.

Most simply, the "new class" consists of men and women with technical or intellectual skills who sell their services and hold jobs—contrasted with an older middle and upper class of men and women with capital who own businesses and an older working class of men and women who have only their labor power to sell. The "new class" is in fact not new, but it has expanded at an extraordinary pace in recent decades and is still growing. Because its members are job-holders, Marxist writers have wondered whether they might not be proletarianized, assimilated into the ranks of the skilled workers. Because they control, manage, and advise other people, conservative writers view them as potential (if currently unreliable) recruits for a new Establishment. Since the "new class" is fairly heterogeneous in character, both these views may be right; or neither. The term does not yet evoke a shared social identity or political position. In neoconservative argument it is used with remarkable freedom, and can be used freely because it isn't connected with any developed political sociology.

Still Steinfels argues persuasively that neoconservative thought is best understood as an ideology for the "new class." It is certainly true that neoconservative writers believe that the "new class" needs an ideology. Its members are *arrivistes*, but they have not arrived by making money, and so they have not been disciplined by the free-for-all of the market. They have no stake in the country, but only in their own persons. They lack understanding and regard for capital. They are as unsure of their own authority as they are of the authority of their predecessors. "Relative to other segments of society," writes Steinfels, "the 'new class' is thin-skinned about legitimacy, high-strung, liable to a 'case of the nerves.'"

Moynihan adds that its members are not aggressive enough in defense of their own interests and of the system within which those interests are pursued. "I would suggest," he told a group of Harvard alumni in 1976, "that a liberal culture does indeed succeed in breeding aggression out of its privileged classes and that after a period in which this enriches the culture, it begins to deplete it." Considering Moynihan himself, a prototypical member of the new class, and his associates in several recent administrations, I don't quite see where the problem of insufficient aggression lies.

The real danger, according to neoconservative writers such as Kristol and Robert Nisbet, is that the "new class" will provide political and social support for a kind of statist egalitarianism. Egged on by radical intellectuals, its members will rally to a "new politics" of leveling, the crucial effect of which will be to enhance the power of the federal bureaucracy, manned by themselves. In other words, they will pursue their own interests (aggressively?). And so they have to be initiated into the complexities of American pluralism. Above all, they have to be taught (through the efforts of foundations like the American Enterprise Institute) to accommodate themselves to the traditional centers of economic power. What the "new class" requires is an ideology that justifies *classes*. It is difficult to doubt, however, that the political practice that goes along with this ideology will be technocratic, elitist, and *dirigiste* in character. The restoration of the bourgeois or of the pre-bourgeois state is not on the neoconservative agenda. What is on the agenda, as Steinfels describes it, is the rule of "policy professionals"—where "professional" means a liberal bureaucrat who is pessimistic about liberation but respects the liberties of the market, who admires local communities and secondary associations but dislikes participatory politics, and who has the strength of mind to enjoy the privileges of his position. And then civility is a creed for the rest of us: teaching a proper respect for our meritocratic betters.

IV

Steinfels obviously thinks civility is more than that. He is a sympathetic critic of neoconservatism—genuinely sympathetic and very much a critic. A highly intelligent Catholic radical, he chooses in this book not to press, indeed barely to put forward, his own position. But he clearly doesn't believe that the alternative to the nervous liberalism of the neoconservatives is a brash and buoyant liberalism. His own view of the present crisis overlaps with theirs; he understands the dangers of

the widespread distrust of institutions among all classes, the dissolution of religious values and the proliferation of cults . . . the anomie and hostility of many inner-city youth, the drift and hedonism of much popular culture, the abandonment of the vulnerable to bureaucratic dependency, the casual amorality of the business world, the retreat from civic consciousness and responsibility. . . .[6]

But he insists that none of these can be dealt with unless one is prepared to examine the "faultlines" of liberal capitalism. This the neoconservatives don't do (Bell, again, is a partial exception). Hence, their concern for "moral culture"—their great strength, according to Steinfels—is vitiated. They argue rightly for the "supporting communities, disciplined thinking and speech, self-restraint, and accepted conventions" that a healthy

[6] *The Neo-Conservatives*, p. 212.

moral culture requires. But they do not tell us, and they cannot, how moral health is ever to be regained, for they have not yet looked unblinkingly at the processes through which it was (or is being) lost.

There is a positive argument that follows from this sort of criticism. Steinfels does not make it, and so I can't tell what form it would take in his hands. This book leaves one waiting for the next. The argument might go something like this. If the old "supporting communities" are in decline or gone forever, then it is necessary to reform them or build new ones. If there are to be new (or renewed) communities, they must have committed members. If marginality and deference are gone too, these members must also be participants, responsible for shaping and sustaining their own institutions. Participation requires a democratic and egalitarian politics—and that is also the only setting, in the modern world, for mutual aid and self-restraint. "The spirit of a commercial people," John Stuart Mill wrote almost a century and a half ago, "will be, we are persuaded, essentially mean and slavish, wherever public spirit is not cultivated by an extensive participation of the people in the business of government in detail. . . ."[7] The argument is as true today as it was when Mill wrote, and far more pressing. Neoconservatism represents the search for an alternative argument, alert to the meanness and slavishness, defensive about commerce, hostile to participation. The search is powerfully motivated and often eloquently expressed, but I do not see how it can succeed.

(1979)

---

[7] "M. de Tocqueville on Democracy in America," in *The Philosophy of John Stuart Mill*, ed. Marshall Cohen (New York, 1961), p. 141.

PART II

The New Left

6

The New Left
and the Old

I

It is not easy to get at the New Left. Already encumbered
with its own myths, hard pressed by the endemic frustrations
and outrages of American society, racially split, infiltrated by
Old Left sectarians, the object of a curious literary cult, it is
no longer the open movement of the early sixties with its
buoyant optimism and transparent passion. Whether anything
at all survives of the radical efflorescence of those years is itself a question. I am going to answer that question in the
affirmative, but only after a rather tortuous description of
what has been a tortuous, though also very short, history.
Rarely in the past has a "new" radicalism been confronted so
quickly with so many impossible choices; rarely has the political resiliency and stamina of the young been so severely
tested. Today, a sense of isolation, an embittered mood, a
dangerous desperation mark many elements of the New Left
like so many scars of battles fought and lost: the collapse of

the civil rights movement, the failure to organize the poor, the continued escalation of the Vietnam War.

The war is perhaps the most important explanation for all that has happened. It is for many of us, and especially for young radicals, a daily humiliation simply to live in the United States while that war is waged in our name. And that humiliation breeds the terrible anger (and the self-hate) and the desire for dramatic "confrontations" that have become characteristic of many student leftists. But there are other reasons, if not better ones, more deeply rooted in the experiences of the past seven years.

II

As a visible political movement, the New Left has its origin in the wave of sympathy and support for the Negro civil rights struggle that swept northern campuses in the early sixties and culminated in Mississippi Summer 1964. But if the militancy of black students is easy enough to account for, that of their white counterparts is not. Black radicalism, even in its most extreme forms, fits admirably into any of a dozen conventional explanations; the white New Leftists are harder to figure out. The struggle for civil rights was less the cause than the occasion for their commitment. Once the call went out, it became clear that many of them had been waiting—but why had they been *waiting?*—and that they had been prepared for political action by something other than the sheer oppressiveness of their surroundings. New Leftists tended to be middle-class students, often at the most prestigious of our universities. Theirs was the radicalism of a generation for whom neither security nor money had ever been a problem. Their parents, by and large, had been children during the worst of the depression, had married and raised families of their own during the war and the postwar boom of the forties, and had rarely managed to convey to their offspring any sense but that of easy expectation. They had outlived, outgrown, or outmaneuvered the various radicalisms of their youth, arriving finally, many of them, at a state of mind that eager sociologists called the end of ideology. They were comfortable, often newly comfortable, and their children inherited from them, in addition to their comforts, only the vaguest idealism, corroded by a new and very strong feeling for the possible pleasures of private life. Yet many of these same children became New Left radicals.

It is a cliché of current political analysis that the New Left grew up as a youthful revolt against the emptiness and hypocrisy of middle-class life. As with most other clichés, there is a truth here, but a truth badly stated. Middle-class life is both interesting and honest enough so long as its discipline serves a real purpose, that is, the pursuit of security and wealth by men and women who possess (and remember having possessed) neither. The radicalism of young people today is not so much a revolt against the emptiness of their parents' lives—for their parents' lives have often been full of struggle, risk taking, and achievement—as against the possible emptiness of their own lives were they simply to take over what their parents have won. For many of them the discipline of professional careers and suburban respectability makes no sense: it will bring them nothing they don't already have. Like every new generation, they want useful and exciting work to do. *But what is the useful and exciting work of the post-affluent generation?* There is a very old "Old Left" answer to this question, to the effect that only when material goods have been won is it possible to pursue moral goods. "First feed the face, and then talk right and wrong." The faces of middle-class America are well fed, so now it is the time to talk. And of course the first thing young people have to say is that the world they would have inherited (and will yet inherit) from their parents is all wrong.

They mean partly that it is wrong that their easy affluence

isn't more widely shared, that in the pursuit of security and wealth so many Americans have been left so far behind. But the specific content of New Left radicalism is not determined by the culture of poverty any more than it is determined by southern black culture—neither of which its leading participants can possibly know—but rather by the culture of plenty. And what New Leftists dislike about the culture of plenty is precisely that controlled efficiency, that careful calculation, that concentration on self and family, that inwardly focused zeal, all of which have been central to the rise of the middle class as a whole and of this or that ethnic group into the middle class, and all of which today's poor will one day emulate. The politics of this culture is largely passive (whatever its conventional moral commitments), marked by the same inward concentration: middle-class Americans surrender almost eagerly the very idea of an active public life, forego the excitements of political action, and seek instead (and *get*) protection and peace of mind at the hands of a benevolent state bureaucracy. The New Left defines itself by opposition: hence its counter-ethos, focused outward, reaching for personal contacts beyond the family circle, emphasizing spontaneity and openness. And hence its counter-politics, demanding a share in the perils and pleasures of power, planning to replace benevolent administration (or certain specified benevolent administrators) with small group democracy and popular participation. It might well be said of most New Leftists that they can afford to be warm, loose, open, and free; that they have time enough and to spare for public activity; and that they have been well trained indeed in all the skills necessary for political participation. But this is no disparagement of their zeal; it merely suggests that their zeal is closely connected, as is everyone else's, to their social position. New Leftism is the politics of a post-affluent class, or of some part of that class, and is probably a politics fully available only to members of that class.

Unfortunately, however, it has only sometimes been possible for young radicals to center their activities in those social areas or to concentrate on those issues where their ideology and experience are directly relevant. Most often they have been driven by the condition of their society and by the moral demands of their age to involve themselves in the life and politics of pre-affluent groups. Thus, some of them have engaged themselves in the black struggle for equality and others in the war (some even in the War) on poverty. And they have sought, as best they could, to apply their ideology and to act out their zeal in radically unfamiliar circumstances. The primary result of this effort is the theory and practice of community organizing, the central theme and the dominant mystique of the New Left today.

III

Community organizing might be described in this way: it is an effort to teach participatory democracy to the poor. Or, less crudely, it is an effort to persuade the poor that they have a great deal to gain through a particular sort, and less to gain through any other sort, of political activity. I should say at once that not many of the poor have been persuaded, and perhaps for good reasons. For the most immediate goals of poor people in the United States today are most unlikely to be won through community organizing in the New Left style. That is not to say that such organizing is of no value, even in the short run, but it is valuable only in so far as it plays into or leads toward the creation of larger organizations—trade unions and political machines—of a sort that New Leftists generally do not regard with favor. The struggles of pre-affluent men and women require for their success two things above all, mass and discipline, and New Left organizing, in part precisely because of its personal intensity and democratic virtue, cannot provide either. The one great advantage of the poor is their number, and that can only be given its proper weight when all the poor people in a given area are mobilized for some concerted action, through union solidarity, for example, or bloc voting, mass demonstrations, and community boycotts. Popular participation obviously plays a significant part in any such mobilization, but so does central (and sometimes charismatic) leadership, an efficient staff, and a widespread willingness to obey commands. Full-scale internal democracy may have to be sacrificed—as it often has been in socialist parties and trade unions in the past—for the sake of the immediate struggle.

It has been one of the achievements of the New Left to remind us (again!) of the full extent of that sacrifice and of the legacy it leaves to the future. That legacy is two-fold: bureaucratic service organizations, centrally controlled, generally benevolent, but unresponsive to popular demands, on the one hand; and passive members with only the dimmest memory of the battles waged in their name, on the other. If the most pressing purposes of the poor are served by this outcome (and they are), surely it is not amiss to suggest that certain broader human purposes are not. This is the burden of radical criticism today and it is the key reason for New Left attempts to organize the poor in other than the obvious and conventional ways. Some New Leftists, of course, argue that the conventional ways won't work, won't bring even the limited gains for which they were designed. That seems to me wrong, and perhaps it would be useful to suggest just how wrong it is by attempting a quick outline of the conventional ways and the limited gains. I mean to point up the precise role and ultimate inadequacy of New Left organizing and, at the same time, the possible truth of the New Left critique. The five stages of political activity that I am going to describe have been derived from the history of the labor movement and of various ethnic groups: I believe they apply also to today's poor in general and to blacks in particular, though perhaps to blacks only with some amendment.

Stage one: passivity—sporadic violence. This is a period of oppression sullenly endured. Poverty is opposed and sometimes overcome only by individual efforts. The poor, whether identified ethnically, racially, or simply economically, constitute what Marx called a class in itself but not for itself. Its members are invisible men; they are treated in effect like things. Occasionally they rebel against this treatment, but the rebellions are formless, without discipline or program, rural or urban *jacqueries.*

Stage two: early mobilization—demonstrations, riots—sectarian activity. Now group consciousness begins to develop and with it there comes a proliferation of (generally tiny) associations of militants who claim to represent the group as a whole and who turn out radical, often imaginative, programs in its name. Sometimes these are secret associations, pledged brotherhoods with blood oaths and an esoteric lingo; sometimes open bands of ideological zealots; sometimes they are made up of home-grown militants; sometimes, as in the case of the New Left today, of missionary radicals. These sectarian clubs really represent nobody, but they do symbolize and help to stir up a new mood of self-assertion, manifest also in demonstrations, strikes, and riots—in which the sectarians play a part, sometimes an important part. None of these, however, can yet be sustained; nor, when they are brutally suppressed, do they leave behind significant organizational residues.

Stage three: high mobilization—political parties and machines, trade unions. Genuinely representative organizations at last appear, usually operating within the political or economic system, challenging its present elites but not necessarily its basic structure. These organizations can be more or less radical in character; their agitators commonly employ a populist rhetoric of one sort or another. The sectarian militants are gradually pushed out of them, however, as large numbers of men and women rush to join, ready now to accept the discipline and share in the hard work necessary to sustain cooperative action. Bloc voting and strikes are typical expressions of the new political competence of previously oppressed and excluded social classes. Both, it should be said, have only limited purposes.

Stage four: partial success—accommodation. The oppressed groups, or a significant number of their members, break into the affluent or near-affluent world, which expands to admit them. Unlike the old aristocracy, the Western middle classes seem capable of infinite expansion. This is true in large part because of the economic growth which they champion, but it is also true because the middle classes have no exclusive style; their way of life can be imitated and sustained at different income levels. Hence rising groups have been able to establish themselves, if not on the peaks of bourgeois wealth and power, then, so to speak, on the slopes—higher or lower. They seize one or another local government, and use its financial resources to help themselves. They win bargaining power in one or another industry and use that to boost wages, establish grievance machinery, and so forth. These are real successes, which should not be denigrated; they are also partial successes, which do not fulfill the programs of stages two and three.

Stage five: demobilization—bureaucracy. Even partial successes have to be defended, but they don't have to be defended by the same kind of organizations that achieved them. The relatively high level of mobilization and action necessary to the achievement now ceases to be necessary. Active participants are largely displaced by competent bureaucrats; open struggle gives way to lobbying and negotiation. Tests of strength still occasionally occur; it is possible to imagine temporary reversions to stage three. But by and large accommodation works; it gives rise to a characteristic passivity, manifest now as privatization, the enjoyment of the limited delights of middle-class society, the rearing of children capable of a new discontent.

This is the long-term process into which New Leftists have

inserted themselves by journeying south or into the slums and ghettos of our northern cities. Their stated purpose is to avoid its likely outcome. They are, after all, the products of that outcome, and so they know or think they know, and even more they feel, how awful it is. I suspect they have some difficulty communicating that sense, even if it does serve to reinforce the natural defensiveness of oppressed and deprived social groups. Has there ever been a myth more generally useful than that of the poor little rich boy, here personified by the young radical from the suburbs who seeks refuge in the slums? But since this young radical is committed to teaching slum dwellers the political skills necessary to escape the slums, and since that escape is widely desired, his position must be extraordinarily ambivalent and painful. For where will the poor go when they escape (either individually or collectively) except into one or another section of middle-class America? This is a difficulty which some New Leftists have resolved by finding, or pretending to find, values among the poor superior to those they knew at home; the poor already have a collective life—a life focused outward to the street and the gang rather than inward to the family—and, above all, a personal looseness and spontaneity which any middle-class American, so it is said, might well envy (and which many do envy). Hence they need freedom and power —to be what they are—rather than bourgeois wealth and security. Possibly a discovery of some moral significance is involved here, even if it is often marred by a perverse sentimentality. But what political conclusions can be drawn from it? On the one hand, the poor cannot win even minimal political power without transforming themselves, not totally, but in important ways. And on the other hand, post-affluent middle-class whites cannot become black and cannot easily become poor. Such parallels as may exist between New Left and ghetto styles are temporary and coincidental, not harbingers of a shared future.

In practice, New Left community organizers move in two

rather different directions, both of which lead them away from the specific content of their own ideology, away from participatory democracy if not from small groups. Some of them —perhaps the best of them—throw themselves into the conventional Old Left work of organizing the poor into unions and political machines, striving for marginal differentiation, but often rediscovering Old Left illusions about the long-term effects of their work. They argue, as Marxists did before them, that this time accommodation will not be possible, this time the organized poor will lead a revolution rather than another invasion of middle-class society. Other New Leftists have pursued the logic of their sentimental identification with the poor as far as it will go. They identify not only with the American poor, but with the poor the world over; they see the ghetto writ large in the Third World; they describe ghetto riots and guerrilla insurrections as if they were the same thing. They extend their commitment at the expense of its efficacy and perhaps because it has had, so far, so little efficacy. And then they eagerly await what they can hardly participate in: an apocalyptic Third World challenge to the America they grew up in. What the American poor make of all this can only be imagined.

New Leftists went into the slums for two reasons: because they were conventional middle-class youth, well-trained and highly competent, with something to teach; because they were unconventional middle-class youth, radically discontented, contemporary *narodniks*, certain that they had something to learn. The two reasons were both good ones, but the tension between them was hard to live with, especially in difficult conditions of daily struggle and danger. What has often (not always) happened, I think, is that middle-class radicals at work in the slums and ghettos have lost confidence in their own talents, above all in the value of their critical faculties and self-discipline, and have become the passive advocates of the going form of slum and ghetto militancy (as of the going form of Third World militancy), whatever its precise content. This is perhaps especially the case with Black Power, which seems so entirely at odds with any authentic New Left ideology, but which few New Leftists would today repudiate. It is also true more generally: the moral and psychic tensions of the encounter in the ghetto, for example, go a long way toward explaining the current New Left view of violence, with its peculiar mix of fascination and fear. Violence is one of the things middle-class radicals learned about among the poor, from the poor themselves and from the oppressors of the poor. The New Left originally was committed to nonviolence, indeed to a special sort of gentleness, openness, personal contact, and co-operation—all of these post-affluent values. America as a whole was and is differently committed, and the politics of personal contact was first transformed into the politics of "confrontation" with all its rhetorical extravagance and misplaced emotion through the experience of community organizing, the encounter with the other America. The young missionary in the slums had endlessly to prove himself in the face of local suspicion and police brutality. Often in proving himself he lost himself, surrendered his special vision and his greatest strengths, and ceased to be useful to the people he had come to help. Among experienced New Leftists, community organizing is said to have a "radicalizing" effect; perhaps it does; it also has an alienating effect, turning middle-class radicals into vicarious guerrillas and Leninist ideologues —neither of these being much-needed sorts of people in America today.

IV

The continued escalation of the Vietnam war has served to aggravate all these tendencies. It overdetermines the New Left thrust toward rage, alienation, self-hate, and ideological rigidity; it produces an apolitical politics in which what seems to be at issue is more often the integrity of the individuals involved than the policy of the state. I don't mean to suggest that the New Left response to the war—there hasn't, of course, been a uniform response—has been irrational or even that it is wrong; I'm not sure what a proper response would be. America these days is infinitely hard on its radicals. All of us have come, however reluctantly, to share Allen Ginsberg's vision: "I saw the best minds of my generation destroyed by madness, starving, hysterical, naked. . . ."[1] Perhaps New Leftists are especially susceptible, and not only because they are—as they undoubtedly are—among the best minds of the new generation; they are especially susceptible also because of their anomalous position in the other America. Their authentic ideology is a response to the special world of affluence, efficiency, and bureaucracy; their authentic politics is one of participation and personal responsibility. But neither this ideology nor this politics provides any adequate means of coping with a brutal, immoral, and seemingly endless war, or with the men who carry on that war. In a peculiar way New Leftism is parasitic on liberalism; it takes off, so to speak, from the peaks of liberal success. When liberals act like the ugliest reactionaries, the New Left is disarmed—capable certainly of the most passionate denunciations, the most outraged expressions of betrayal and contempt, all of this well deserved, but utterly incapable of effective action and sometimes even of coherent thought. Young radicals have talked a great deal about building a mass movement against the war, but the techniques they have adopted (and which are probably most appropriate to them) are ill suited to that goal. They tend instead to create enclaves of moral men and women in an ugly and insane world, men and women whose mark is not their commitment of middle-class competence and discipline to a cause, but rather their willingness to "put their bodies on the line." But what else ought they to do? It is not as if anyone

---

[1] *Howl and Other Poems* (San Francisco, 1956), p. 9.

had succeeded in building an antiwar movement distinguished by its size, its unity, or its effectiveness, which New Leftists might join or where they might work part time even while maintaining their enclave. In the absence of a meaningful liberalism the burden of moral expression has fallen disproportionately on them, and they have both assumed that burden and suffered from it.

At the same time, the war has intensified an ideological development that began in the slums. It has led New Leftists to see the affluent world from which they came as a world literally dependent upon the systematic exploitation of masses of people at home and abroad. The theory of imperialism is today more widely accepted in the United States than at any time in the recent past (with some reason, after all), and this means that one of the most important Old Left ideologies has become a prevalent New Left ideology. The more post-affluent radicals are driven to confront the painful realities of the pre-affluent world, the more such old ideologies are likely to gain ground. For they have, whatever their intellectual cogency, a certain moral relevance to the social conditions in which they were bred. Not necessarily such a relevance as will make them useful guides to political action: their effect is more often to make possible plausible explanations for the failure of whatever action is undertaken, and then to provide plausible reasons for a withdrawal from a corrupted America into sectarian rectitude. So the New Left inherits not the victories but the defeats of the past, and in so far as it makes its peace with that inheritance, begins to transform itself from a moral enclave into a political sect. That transformation has not yet gone very far; the New Left still possesses many of its original qualities. Whether these can survive the Vietnam War, however, is a hard question.

So long as that war continues, opposition to it is bound to grow, and the New Left forms of that opposition—most crucially draft resistance—are also going to grow. Draft resistance is not likely to end the war; nor is the New Left, having carried personal responsibility to such a pitch, likely to function usefully in whatever more moderate antiwar movement the country may eventually produce. Too many New Leftists have come to doubt the very capacity of the country to offer a politics they might support. The best that can be hoped for is that draft resistance will shame liberals into a less pusillanimous opposition to the war. Then the moral fellowship that it generates in the New Left will not be so totally alienated from American life as to be incapable of functioning creatively in the postwar world.

V

I have left until last any consideration of what New Left creativity might be like at home, so to speak, in its proper environment. The university best serves to represent that environment; it is the first, and still the only, home ground of the New Left, the major center of its activity even during the most intense period of community organizing. And lately a certain return to the campus by New Left organizers has become noticeable; it occurs most immediately because the university is the only satisfactory base for antiwar work. The slums, unhappily, are not a good base at all. But there is also a more general reason: the modern university is pre-eminently an institution of the post-affluent world, inhabited by middle-class students, run by liberal administrators, interlocked in a great variety of ways with governmental and corporate bureaucracies. Ultimately, of course, New Leftists must find appropriate methods of challenging the government and the corporations. They are not wrong to imagine, however, that campus agitation is more than a way of preparing for that future work; it is itself a beginning, the first confrontation with that special modernity it is their purpose to transform.

Student radicalism has taken many forms in the past year or two: resistance, the attack upon university "complicity" in the Vietnam War, the fight for academic freedom and for student power, the effort to promote a "radical education"—the series is not without internal tension. I am going to concentrate on the question of student power, since it brings us immediately into touch with what I have been calling the authentic ideology of the New Left.

Student power is sometimes viewed as a campus corollary to Black Power, largely because of the myth (more prevalent among New Leftists than it ought to be, given their general dislike for self-serving arguments) that students constitute an oppressed class. They do not. They belong, as most of them know and as everybody else knows, to a privileged class. But as is customary in such classes, they are asked to pay a price for their privileges. They are asked to accept the social system that designates them as privileged members, to learn its discipline and the special skills it requires, and to obey its rulers until the time comes when they, or some of them, become rulers themselves. In the context of the university this means that students are asked to keep quiet and pay attention, and to dissipate their conventionally rebellious energies in conventionally approved ways. The demand for student power represents, first of all, a refusal to do these things. It is an argument that the character of the social system is or ought to be an issue for its privileged as well as for its unprivileged members. The students rightly see the university as microcosm of American society at its most modern, an embodiment of the system and not merely one of its training camps; and so when they rebel they are not so much refusing to be trained as demanding the right to experiment, here and now, with new and possibly better social arrangements. And the demand for student power is also, secondly, an argument as to the best experiment, an argument for radical democracy.

There are difficulties here above and beyond the predictable resistance of university administrators. The university is not, or not only, a community of equals; it is also a community in which a particular set of people, who are also older, lay claim to special competence in the common work. And though the power they derive from that claim may well be as corrupting as the power of a southern sheriff (or an American president), it does not make a great deal of sense simply to equate the university power structure with all other structures. Nor is it useful (though it is conceivably possible) to set up confrontations on the campus that resemble those of Selma, Detroit, or Dak To. All sorts of fantasies are acted out on college campuses, from the revolutionary indignation of the sailors in the mess of the battleship *Potemkin* to the storming of the Pentagon, but the confusion of any of these with New Left *politics* is a plain disaster for the New Left.

That having been said, and all necessary allowances made for the pressures of the war and the militancy they breed, there remains a great deal that is both sensible and attractive in the idea of student power. Students who press that idea in their communities are performing at least three valuable services. First, they are protesting against the increasing depersonalization of the university and of education itself—a fact of our common life that requires the continual testimony of those who are hurt by it, so easily does it become a cliché, the subject of occasional homilies and everyday acceptance. Depersonalization is only partly caused by the enormous growth in the number of university students. It is much more importantly the product of a series of ideologically governed decisions as to how to educate these students, and where, and with what degree of social expenditure. By and large, I think, these decisions have been bad ones, revealing the fundamental shoddiness of our welfare state and the dishonesty of our commitment, as a society, to humane values. Taken together, these decisions have required a very narrow kind of efficiency from university administrators, an efficiency not unlike that required—for other though perhaps related reasons —from corporate managers and government bureaucrats. And whatever its value in business and government, this efficiency is deadly in schools, for it goes some way toward depriving the students of those close contacts with exemplary adults and intense relations with one another that are so important to the business of growing up. Goes *some* way because, of course, it does not entirely succeed; the New Left itself is an example of the opposition it breeds.

Second, campus radicalism is a protest against the increasing interconnections of university administrations and faculties with business and government. This is not only a matter of "complicity" in the Vietnam War; it is more importantly a matter of the progressive disappearance of any moral center around which an autonomous university might grow. Student power is one of the more impressive instances of the New Left defense of local autonomy and small-group activism. When students derisively call the modern university a "service station," they are not objecting to the idea that a school must provide concrete and socially useful skills—it must—but they are denying that that is all a school must do. A service station is not an ethical world; life within it may have advantages, but it has no value. Indeed, university life is drained of value when its potential leaders—teachers above all—turn outward to face the state, where, they imply, all the action is. The loss of local value and local action—the integration of small communities into larger worlds centered *there* but never *here*— this again is a long-term tendency of modern society, but it is surely not inappropriate to suggest that the university is a good place to begin to resist. That resistance, of course, cannot take the form of a demand that universities become service stations of a new sort, adjusted to the needs of lower- rather than upper-class constituencies, of oppositional rather than established politics. Students can serve whomever they please; so can professors (in their "leisure" time); but an autonomous university must be committed above all to the education of its own members.

Finally, the question of power. Student power is not an end in itself; even some minimal degree of participation in local decision making would have the immediate effects of improving the quality of campus relations and of enlivening community activities. But there is an independent argument to be made for such participation; it is the classic argument for democratic government. Democracy in a university setting is necessarily subject to significant limits; its primary form is faculty and not student self-government. But there are a variety of ways in which students can be involved—most directly in the running of their own dormitories and extracurricular activities. And there is one sphere of educational activity, vitally important, and entirely in their own hands: the sphere of the study group, the political club with its theoretical journal, the extracurricular course. Here the Old Left scored some notable triumphs, seizing the intellectual initiative at a number of key universities in the 1930s. The New Left has been less successful, perhaps because students today feel pressed toward militant postures by the war, perhaps because they have less sense than their predecessors did of standing on intellectual ground, where ideas and theories matter. Or perhaps they have learned too quickly to think of themselves as consumers of education, who cannot take the lead, who must receive their rights and privileges from the hands of the authorities. In any case, the combination of their passivity and their anger is a great misfortune, and some form of student power is probably an appropriate remedy. Students have a lot to learn, but they are not politically or intellectually incompetent. The failure to involve them actively in university life, their failure to involve themselves: for all this, there will one day be, as the saying goes, hell to pay.

VI

One form that hell takes is the real incompetence, passivity, and moral lethargy of so many of the citizens of the modern state—all of these qualities painfully in evidence in the United States today, when citizens are forced to face the reality of an immoral war. Against all this the New Left is impotent and its members hard pressed to maintain their sanity. It is not, and is most unlikely to become, a mass movement, and that is why it is of so little help to those Americans for whom a mass movement is still an urgent necessity. It is also not a revolutionary movement—whatever the rhetoric of some of its leaders—for it aims at what are really incremental victories against the presently overwhelming character of governmental authority and popular inaction, victories to be won first in this community, then in that one. The university is obviously only one such community, conceivably the easiest one, and if the young radicals are to "grow up political" (and not absurd), they will have to find a way of working in other places. This they have not yet done, chiefly because the places they have chosen or, better, been driven to choose are not the places where they are likely to be effective. They are trapped between the America they know and the other America, alienated from the one, frustrated in the other. They live in a kind of limbo; their politics for the moment is a dance: two steps forward, two steps back.

The crucial condition of their future success, indeed, of their very survival as a *new* political movement, is that things start moving again in the other America. That would be a motion to which they could contribute, if only marginally, and it would carry them (and others too) forward into a world where they could begin to act out their authentic politics. Until that happens, however, the confrontations of New Leftists with the post-affluent world, with their own America, are bound to be deflected and their significance and value distorted, as they already have been to some extent, in ways I have tried to describe. A politics of protest against the bureaucratic impersonality and omnicompetence of the welfare state requires for its success a fully realized or more fully realized welfare state. And that means that Americans must earn the right to have a *New* Left by completing the work of the Old.

(1967, 1968)

127

7

A Day in the Life of a Socialist Citizen

Imagine a day in the life of a socialist citizen. He hunts in the morning, fishes in the afternoon, rears cattle in the evening, and plays the critic after dinner. Yet he is neither hunter, fisherman, shepherd, nor critic; tomorrow he may select another set of activities, just as he pleases. This is the delightful portrait that Marx sketches in *The German Ideology* as part of a polemic against the division of labor.[1] Socialists since have worried that it is not economically feasible; perhaps it is not. But there is another difficulty that I want to consider: that is, the curiously apolitical character of the citizen Marx describes. Certain crucial features of socialist life have been omitted altogether.

In light of the contemporary interest in participatory democracy, Marx's sketch needs to be elaborated. Before hunting in the morning, this unalienated man of the future is likely

[1] Karl Marx and Friedrich Engels, *The German Ideology*, ed. R. Pascal (New York, 1947), p. 22.

to attend a meeting of the Council on Animal Life, where he will be required to vote on important matters relating to the stocking of the forests. The meeting will probably not end much before noon, for among the many-sided citizens there will always be a lively interest even in highly technical problems. Immediately after lunch, a special session of the Fishermen's Council will be called to protest the maximum catch recently voted by the Regional Planning Commission, and the Marxist man will participate eagerly in these debates, even postponing a scheduled discussion of some contradictory theses on cattle rearing. Indeed, he will probably love argument far better than hunting, fishing, *or* rearing cattle. The debates will go on so long that the citizens will have to rush through dinner in order to assume their role as critics. Then off they will go to meetings of study groups, clubs, editorial boards, and political parties where criticism will be carried on long into the night.

Oscar Wilde is supposed to have said that socialism would take too many evenings. This is, it seems to me, one of the most significant criticisms of socialist theory that has ever been made. The fanciful sketch above is only intended to suggest its possible truth. Socialism's great appeal is the prospect it holds out for the development of human capacities. An enormous growth of creative talent, a new and unprecedented variety of expression, a wild proliferation of sects, associations, schools, parties: this will be the flowering of the future society. But underlying this new individualism and exciting group life must be a broad, self-governing community of equals. A powerful figure looms behind Marx's hunter, fisherman, shepherd, and critic: the busy citizen attending his endless meetings. "Society regulates the general production," Marx writes, "and thus makes it possible for me to do one thing today and another tomorrow."[2] If society is not to become an alien and dangerous force, however, the citizens cannot accept its regu-

---

[2] *German Ideology*, p. 22.

lation and gratefully do what they please. They must participate in social regulation; they must be social men and women,
organizing and planning their own fulfillment in spontaneous
activity. The purpose of Wilde's objection is to suggest that
just this self-regulation is incompatible with spontaneity, that
the requirements of citizenship are incompatible with the
freedom of hunter, fisherman, and so on.

Politics itself, of course, can be a spontaneous activity, freely chosen by those men and women who enjoy it and to whose
talents a meeting is so much exercise. But this is very unlikely
to be true of all men and women all the time—even if one
were to admit what seems plausible enough: that political
life is more intrinsic to human nature than is hunting and
cattle rearing or even (to drop Marx's rural imagery) art or
music. "Too many evenings" is a shorthand phrase that describes something more than the sometimes tedious, sometimes exciting business of resolutions and debates. It suggests
also that socialism and participatory democracy will depend
upon, and hence require, an extraordinary willingness to attend meetings, and a public spirit and sense of responsibility
that will make attendance dependable and activity consistent
and sustained. None of this can rest for any long period of
time or among any substantial group of people upon spontaneous interest. Nor does it seem possible that spontaneity will
flourish above and beyond the routines of social regulation.

Self-government is a very demanding and time-consuming
business, and when it is extended from political to economic
and cultural life, and when the organs of government are decentralized so as to maximize participation, it will inevitably
become more demanding still. Ultimately, it may well require
almost continuous activity, and life will become a succession
of meetings. When will there be time for the cultivation of
personal creativity or the free association of like-minded
friends? In the world of the meeting, when will there be time
for the *tête-à-tête*?

I suppose there will always be time for the *tête-à-tête*. Men
and women will secretly plan love affairs even while public

business is being transacted. But Wilde's objection is not silly. The idea of citizenship on the Left has always been overwhelming, suggesting a positive frenzy of activity and often involving the repression of all feelings except political ones. Its character can best be examined in the work of Rousseau, from whom socialists and, more recently, New Leftists directly or indirectly inherited it. In order to guarantee public-spiritedness and political participation, and as a part of his critique of bourgeois egotism, Rousseau systematically denigrated the value of private life:

The better the constitution of a state is, the more do public affairs encroach on private in the minds of the citizens. Private affairs are even of much less importance, because the aggregate of the common happiness furnishes a greater proportion of that of each individual, so that there is less for him to seek in particular cares.[3]

Rousseau might well have written these lines out of a deep awareness that private life will not, in fact, bear the great weight that bourgeois society places upon it. We need, beyond our families and jobs, a public world where purposes are shared and cooperative activity is possible. More likely, however, he wrote them because he believed that cooperative activity could not be sustained unless private life were radically repressed, if not altogether eradicated. His citizen does not participate in social regulation as one part of a round of activities. Social regulation is his entire life. Rousseau develops his own critique of the division of labor by absorbing all human activities into the idea of citizenship: "Citizens," he wrote, "are neither lawyers, nor soldiers, nor priests by profession; they perform all these functions as a matter of duty."[4] *As a matter of duty:* here is the key to the character of that

[3] Jean Jacques Rousseau, *The Social Contract*, trans. G. D. H. Cole (New York, 1950), bk. III, chap. 15, p. 93.
[4] Jean Jacques Rousseau, *Considerations on the Government of Poland*, in *Political Writings*, trans. Frederick Watkins (Edinburgh, 1953), p. 220.

patriotic, responsible, energetic man who has figured also in socialist thought, but always in the guise of a new man, freely exercising his human powers.

It is probably more realistic to see the citizen as the product of collective repression and self-discipline. He is, above all, *dutiful,* and this is only possible if he has triumphed over egotism and impulse in his own personality. He embodies what political theorists have called "republican virtue"—that means, he puts the common good, the success of the movement, the safety of the community, above his own delight or well-being, *always.* To symbolize his virtue, perhaps, he adopts an ascetic style and gives up every sort of self-decoration: he wears sans-culottes or unpressed khakis. More important, he foregoes a conventional career for the profession of politics; he commits himself entirely. It is an act of the most extreme devotion. Now, how is such a man produced? What kind of conversion is necessary? Or what kind of rigorous training?

Rousseau set out to create virtuous citizens, and the means he chose are very old in the history of republicanism: an authoritarian family, a rigid sexual code, censorship of the arts, sumptuary laws, mutual surveillance, the systematic indoctrination of children. All these have been associated historically (at least until recent times) not with tyrannical but with republican regimes: Greece and Rome, the Swiss Protestant city-states, the first French republic. Tyrannies and oligarchies, Rousseau argued, might tolerate or even encourage license, for the effect of sexual indulgence, artistic freedom, extravagant self-decoration, and privacy itself was to corrupt men and women and turn them away from public life, leaving government to the few. Self-government requires self-control: it is one of the oldest arguments in the history of political thought.[5]

[5] It is sympathetically restated by Alan Bloom in his introduction to Rousseau's *Letter to M. D'Alembert on the Theatre,* in *Politics and the Arts* (Glencoe, Ill., 1960), pp. xv–xxxviii.

If that argument is true, it may mean that self-government also leaves government to the few. At least, this may be so if we reject the disciplinary or coercive features of Rousseau's republicanism and insist that citizens always have the right to choose between participation and passivity. Their obligations follow from their choices and do not precede them, so the state cannot impose one or the other choice; it cannot force the citizens to be self-governing men and women. Then only those citizens will be activists who volunteer for action. How many will that be? How many of the people you and I know? How many ought they to be? Certainly no radical movement or socialist society is possible without those ever-ready participants, who "fly," as Rousseau said, "to the public assemblies."[6] Radicalism and socialism make political activity for the first time an option for all those who relish it and a duty— sometimes—even for those who do not. But what a suffocating sense of responsibility, what a plethora of virtue would be necessary to sustain the participation of everybody all the time! How exhausting it would be! Surely there is something to be said for the irresponsible nonparticipant and something also for the part-time activist, the half-virtuous man (and the most scorned among the militants), who appears and disappears, thinking of Marx and then of his dinner? The very least that can be said is that these people, unlike the poor, will always be with us.

We can assume that a great many citizens, in the best of societies, will do all they can to avoid what Melvin Tumin has called "the merciless masochism of community-minded and self-regulating men and women."[7] While the necessary meetings go on and on, they will take long walks, play with their children, paint pictures, make love, and watch television. They will attend sometimes, when their interests are directly at stake or when they feel like it. But they will not make the full-

---

[6] *Social Contract*, bk. III, chap. 15, p. 93.
[7] Melvin Tumin, "Comment on Papers by Riesman, Sills, and Tax," in *Human Organization* 18 (Spring 1959): 28.

scale commitment necessary for socialism or participatory democracy. How are these people to be represented at the meetings? What are their rights? These are not only problems of the future, when popular participation has finally been established as the core of political and economic life. They come up in every radical movement; they are the stuff of contemporary controversy.

Many people feel that they ought to join this or that political movement; they do join; they contribute time and energy —but unequally. Some make a full-time commitment; they work every minute; the movement becomes their whole life, and they often come to disbelieve in the moral validity of life outside. Others are established outside, solidly or precariously; they snatch hours and sometimes days; they harry their families and skimp on their jobs, but yet cannot make it to every meeting. Still others attend scarcely any meetings at all; they work hard but occasionally; they show up, perhaps, at critical moments, then they are gone. These last two groups make up the majority of the people available to the movement (any movement), just as they will make up the majority of the citizens of any socialist society. Radical politics radically increases the amount and intensity of political participation, but it does not (and probably ought not) break through the limits imposed on republican virtue by the inevitable pluralism of commitments, the terrible shortage of time, and the day-to-day hedonism of ordinary men and women.

Under these circumstances, words like citizenship and participation may actually describe the enfranchisement of only a part, and not necessarily a large part, of the movement or the community. Participatory democracy means the sharing of power among the activists. Socialism means the rule of the people with the most evenings to spare. Both imply, of course, an injunction to the others: join us, come to the meetings, participate! Sometimes young radicals sound very much like old Christians, demanding the severance of every tie for the sake of politics. "How many Christian women are there," John

Calvin once wrote, "who are held captive by their children!"[8] How many "community people" miss meetings because of their families! But there is nothing to be done. Ardent democrats have sometimes urged that citizens be legally required to vote: that is possible, though the device is not attractive. Requiring people to attend meetings, to join in discussions, to govern themselves: that is not possible, at least not in a free society. And if they do not govern themselves, they will, willy-nilly, be governed by their activist fellows. The apathetic, the occasional enthusiasts, the part-time workers: all of them will be ruled by full-timers, militants, and professionals.

But if only some citizens participate in political life, it is essential that they always remember and be regularly reminded that they are . . . only some. This is not easy to arrange. The militant in the movement, for example, does not represent anybody; it is his great virtue that he is self-chosen, a volunteer. But since he sacrifices so much for his fellowmen, he readily persuades himself that he is acting in their name. He takes their failure to put in an appearance only as a token of their oppression. He is certain he is their agent, or rather, the agent of their liberation. He is not in any simple sense wrong. The small numbers of participating citizens in the United States today, the widespread fearfulness, the sense of impotence and irrelevance: all these are signs of social sickness. Self-government is an important human function, an exercise of significant talents and energies, and the sense of power and responsibility it brings is enormously healthy. A certain amount of commitment and discipline, of not-quite-merciless masochism, is socially desirable and efforts to evoke it are socially justifiable.

But many of the people who stay away from meetings do

---

[8] John Calvin, *Letters of John Calvin*, ed. Jules Bonnet, trans. David Constable (Edinburgh, 1855), vol. 1, p. 371. Of all alternate communities, the family is clearly the greatest danger to the movement and the state. That is not only because of the force of familial loyalty, but also because the family is a place of retreat from political battles: we go home to rest, to sleep.

so for reasons that the militants do not understand or will not acknowledge. They stay away not because they are beaten, afraid, uneducated, lacking confidence and skills (though these are often important reasons), but because they have made other commitments; they have found ways to cope short of politics; they have created viable subcultures even in an oppressive world. They may lend passive support to the movement and help out occasionally, but they will not work, nor are their needs and aspirations in any sense embodied by the militants who will.

The militants represent themselves. If the movement is to be democratic, the others must *be represented*. The same thing will be true in any future socialist society: participatory democracy has to be paralleled by representative democracy. I am not sure precisely how to adjust the two; I am sure that they have to be adjusted. Somehow power must be distributed, as it is not today, to groups of active and interested citizens, but these citizens must themselves be made responsible to a larger electorate (the membership, that is, of the state, movement, union, or party). Nothing is more important than that responsibility; without it we will only get one or another sort of activist or *apparatchik* tyranny. And that we have already.

Nonparticipants have rights; it is one of the dangers of participatory democracy that it would fail to provide any effective protection for these rights. But nonparticipants also have functions; it is another danger that these would not be sufficiently valued. For many people in America today, politics is something to watch, an exciting spectacle, and there exists between the activists and the others something of the relation of actor and audience. Now for any democrat this is an unsatisfactory relation. We rightly resent the way actors play upon and manipulate the feelings of their audiences. We dislike the aura of magic and mystification contrived at on stage. We would prefer politics to be like the new drama with its alienation effects and its audience participation. That is fair enough.

But even the new drama requires its audience, and we ought not to forget that audiences can be critical as well as admiring, enlightened as well as mystified. More important, political actors, like actors in the theater, need the control and tension imposed by audiences, the knowledge that tomorrow the reviews will appear, tomorrow people will come or not come to watch their performance. Too often, of course, the reviews are favorable and the audiences come. That is because of the various sorts of collusion that commonly develop between small and co-opted cliques of actors and critics. But in an entirely free society, there would be many more political actors and critics than ever before, and they would, presumably, be self-chosen. Not only the participants, but also the nonparticipants, would come into their own. Alongside the democratic politics of shared work and perpetual activism, there would arise the open and leisurely culture of part-time work, criticism, second-guessing, and burlesque. And into this culture might well be drawn many of the alienated citizens of today. The modes of criticism will become the forms of their participation and their involvement in the drama the measure of their responsibility.

It would be a great mistake to underestimate the importance of criticism as a kind of politics, even if the critics are not always marked, as they will not be, by "republican virtue." It is far more important in the political arena than in the theater. For activists and professionals in the movement or the state do not simply contrive effects; their work has more palpable results. Their policies touch us all in material ways, whether we go or do not go to the meetings. Indeed, those who do not go may well turn out to be more effective critics than those who do: no one who was one of its "first-guessers" can usefully second-guess a decision. That is why the best critics in a liberal society are citizens-out-of-office. In a radically democratic society they would be people who stay away from meetings, perhaps for months at a time, and only then discover that something outrageous has been perpetrated that must be mocked or protested. The proper response to such protests is not to tell the laggard citizens that they should have been active these past many months, not to nag them to do work that they do not enjoy and in any case will not do well, but to listen to what they have to say. After all, what would democratic politics be like without its kibitzers?

(1968)

8

Violence: The Police, the Militants, and the Rest of Us

I

Politics is first of all the art of minimizing and controlling violence. It is a taxing and morally dangerous art because violence itself and the threat of violence are two of its instruments. These can be put to use with almost equal ease by the public authorities and by private men and women, but almost always more massively and more effectively by the authorities. So it is best to begin by worrying about them.

War, riot control, law enforcement, "maintenance of order," punishment: all these can involve violence. The word itself calls to mind (to my mind, at least) a charging phalanx of helmeted police, an image fixed, I suppose, as much by the media as by recent events. It is by no means an accurate picture of what it is sometimes said to represent: the state stripped naked. But it does make it impossible to pretend that "the use of physical force to inflict injury" (the dictionary definition of violent behavior) is something other than violent whenever the users are uniformed officials, and it is equally impossible to argue that official violence is somehow presumptively legitimate. Even the justification of punishment, the most ordinary form of official violence, is uncertain and much disputed among philosophers. The intermittent violence of everyday law enforcement is more commonly questioned, especially since historians began to compile the record of police and militia brutality against labor organizers, radical agitators, ethnic and racial minorities. And just as the state's use of violence is often cruel, needless, and illegitimate, so the state possesses no monopoly on legitimate use. We deny its monopoly every time we assert the right of self-defense or of collective resistance to tyranny.

What the state does have and should have is control—in this country it has exclusive control—of the means of massive violence, both human and material: the army and police, planes, tanks, artillery, and so on. Only the ownership of guns is legally the right of individuals in the United States, and this is a right increasingly questioned—for very good reasons. The case for state control of the means of massive violence is so widely accepted that I need not review it here. I have to stress instead that the existence of such control, however justified, also establishes the state as a threat to us all. The threat is simply that violence will be used, as it has been so often in the past, not to protect everyone's rights or to enforce policies democratically agreed upon, but to defend the privileges of some few of us or to interfere with or repress democratic liberties. Obviously, no guarantees are possible here. Official violence is like unofficial violence, at least in this regard: it too inflicts injury, and it needs (in every case) to be justified. From the time of Thomas Hobbes on, the most successful efforts to justify the use of violence have undoubtedly been undertaken on behalf of the state. But these efforts do not free us from the burden of asking ourselves, over and over again, what are the prospects for the use of military and police power? Insofar as we defend state control, we have an obligation to watch over it.

Recently, however, many of us have focused, except when talking about the Vietnam war, on unofficial violence. Urban riots, street confrontations, terrorism: these claimed our attention, even if only our critical attention. Confrontation and terror suggested a *politics* of unofficial violence toward which every leftist and liberal had to adopt a position. That was not hard to do, and no one deserves any special credit for setting himself against it, though there is some discredit to be distributed among those who patronized unofficial violence or apologized for it. "Nothing has ever been accomplished without violence"—this was the peculiarly debased slogan of the apologists, suggesting that it was not *only* in order to make omelets that one had to break eggs. It was necessary also in order to show contempt for the art of cooking, to leave the kitchen as messy as possible, to revenge oneself against bourgeois omelet-lovers, and so on. The only trouble is that all this can be accomplished and no one satisfied. And that is true as well of confrontation and terror: they represent a dead end of meaningless militancy. Nor has it taken very long to explore their farthest reaches.

But something happened in the last years of the 1960s that the militants and the rest of us ought to be concerned about. Police power was significantly strengthened and government surveillance over political activity widely extended. That is the usual outcome of unofficial violence, but just what it means for our immediate future is not easy to figure out.

It doesn't mean what we are sometimes told by both radical ideologists and their critics: that the country has been dangerously polarized or that we are in the midst of or on the verge of fascist repression. Instead of polarizing the country, the militants and terrorists have succeeded only in isolating themselves; they are cut off from any hope of a political base or of large-scale (even small-scale) popular support. The country as a whole seems less divided than it was in 1968—I'm not sure how one ought to feel about that. And the repression of the far Left since that time has been less organized and thorough (and the repression of terrorists considerably less efficient) than most of us expected, given the character of the national administration and all the opportunities it has had. There is, nevertheless, a great deal to worry about: above all, the apparent readiness of the police themselves to engage in confrontation politics; also the nighttime police raids; the shooting down of student demonstrators; the large number of men and women under indictment, often for offenses whose precise character is very unclear; the general and apparently not unrealistic sense of being watched and spied upon. Yet all this does not add up to a countrywide war against the Left. Most of the incidents that outrage us seem rather to be examples of local zealotry, reflecting the continuing and not the new weakness of liberal (and, needless to say, of far Left) politics in many parts of the country.

Nationally, however, things are not the same. Despite the Left's romance with localist ideologies, it is protected largely by national forces and would be better protected were this country's politics still more "nationalized." It seems certain that no one would have died at either Kent State or Jackson State had the Ohio National Guard or the Mississippi State Police been under effective federal control. It is significant also that the least bloody of our ghetto riots occurred in Washington, D.C., and again, that federal courts often overturn the convictions won by local prosecutors. At the present time, and for the foreseeable future, massive repression simply does not have the support of national elites, nor can it be worked through national institutions. The reports of one presidential commission after another—they must reflect *someone's* opinion—point as clearly as one could wish to the absence of any ideology of repression at the national level. Nor does the political campaign for law and order seem likely to change any of this. The 1970 elections suggest instead that far Left violence has not been exploited so as to enlarge the base of conservative support. Given all this, talk of an American fascism seems only a particularly gruesome (and for some people, I suppose, entertaining) escape from the need to face honestly the extent of far Left failure—and the real, if more limited, dangers that failure always brings.

But there is escapism also on what might be called the "near Left"—a term I take to include not only democratic socialists, but also many liberals, even those who apologize for far Left violence, since they are apologists at such a distance.

We have been terribly reluctant to face the unpleasant but critically important question: *just what sort of law enforcement do we want?* For example:

Are we in favor of a national police force or have we opted definitively for local control? A national force would surely be more liberal (and better disciplined) than many local forces; at least it would be more liberal right now; but it would also give to the central government a kind of power that liberals and conservatives alike have always warned against.

Do we want the National Guard abolished, or its members more widely recruited and better trained? Many of the most disgraceful incidents in the long history of official violence are the work of the National Guard; yet at the same time it is the closest thing we have to a citizens' militia. (But after reading the polls on the work of the Guard at Kent State, we may be soured on citizens.)

How do we feel about the many local budgetary decisions that are providing more and more money for the police? Like other municipal employees, they have been badly underpaid, but have they also been underequipped? The Kerner Commission said so, but warned at the same time against supplying policemen with automatic rifles and machine guns. And what about mace and helicopters and other devices tested in Vietnam?

What do we think of the infiltration of police spies into far Left organizations? Into terrorist groups? Is there some line to be drawn between people who can be spied upon and people (like us) who cannot be?

At what point do we want police called onto a university campus or sent into battle against political demonstrators or rioters? And then, do we want them to enforce the law (and collect people for punishment) or simply to keep the peace as best they can?

Do any of us really think that policemen are "pigs"? And if anyone does believe that, will he refrain from calling on them when faced with a riot or a bomb threat?

I don't want to suggest by any of these questions that we should be prepared to accept the erosion of civil liberties. The libertarian struggles of the past decade, however, have been largely concerned with the establishment of judicial restraints on police action. These will certainly need to be defended; they are already under attack. I mean to call attention here to the problem of political control. For it is clear that important political decisions have been and will continue to be made by police officials in areas where the courts cannot or will not interfere. They exercise discretion as to what laws they will enforce, with what degree of violence, in what parts of the city, and so on. In some cities, the extent of political control over such decisions is as unclear as is the extent of federal control over national defense. The reasons for the confusion are the same too: larger budgets, advanced weaponry, the freedom that necessarily exists under crisis conditions, and the emergence under these same conditions of the police and the military as political forces in their own right. All this leaves us ignorant of the responsibility (or lack of it) of elected officials. Thus, the war against the Black Panthers looks sometimes like a political campaign (national or, more likely, local) and sometimes like a police crusade. The second would be more understandable, but also more frightening.

The standard Left response is to attack police brutality while insisting that the proper focus of political energy is on those social problems that force public officials to rely on the police. The same response is standard in international affairs, where leftists generally see in this or that instance of military sadism only the necessary result of a failure to solve the problem of imperialism, capitalism, or war itself. But the police, like the army, are likely to be with us for a long time. It is simply not the case that police powers will cease to be necessary and dangerous once a beginning is made toward solving whatever social problems lie at the roots of our present discontents; nor is it certain that all our discontent has roots that governmental problem solving can reach. The near Left must face up to the implications of its attack on police brutality: we want the police to act differently, but not to go away. I suppose what we need is a domestic version of the theory of just war—a theory of "civil law enforcement," the appropriate parallel for civil disobedience, but necessary even when disobedience is uncivil.

What kind of a police force do we (of the near Left) want? Surely a force that is, first, carefully selected, well trained, and highly disciplined; these features alone would considerably reduce the quantity of official violence. They would reduce it, that is, to the precise quantity intended by whoever commanded the police and thereby make their responsibility a great deal clearer. Second, a police force that is well paid: for the increasingly narrow social base from which police are recruited plays havoc with law enforcement. Part of what happens when police come onto a campus is simply class war—to be sure of a curious sort, with radical students pretending to represent and the police more nearly representing the working class. It is not a formula for minimal violence.

Third, we want or should want a force committed to a fairly rigid professional and legal code (a code, obviously, that requires police to minimize violence). And fourth, a force answerable, within the limits set by that code, to elected officials. These last two requirements are not necessarily consistent or not easily maintained together: there is a tension (in the public school system too) between professional commitment and political control. But that is a tension best left unresolved. It's one of the more peculiar notions of the far Left and of many of their liberal apologists that a community police would be better than what we have now. In fact, we have our greatest difficulties precisely with policemen who are little more than ordinary folks with guns. Since it is utopian to hope for an American police force without guns, we must insist that armed men and women be carefully trained. And the idea of training implies the existence of standards.

I am inclined to think, perhaps naively, that a police force better trained and controlled would be less available (in the United States today) as an instrument of political repression. Clearly, it would be a purer instrument, its use less violent and more efficient at whatever task it was set, and so its effects more strictly limited to its legally appointed task. The worst effects of police power as we have recently known it (historically too, so far as I can discover) are those that stem from what might be called its "incidental" uses. The police are most repressive not when they are used to prevent or control illegal political activities, but when allowed or encouraged to *punish* those activities—most often by beating up (or killing) the activists. That is not what they are officially told to do, nor is it what they need to do to accomplish their legal purpose. Police violence obviously serves other purposes, but it also expresses ethnic, social, and political feelings—private feelings that ought to be repressed by professionals doing a job. That's a good kind of repression, and it would make the police less repressive.

At the same time, "incidental" violence is exactly the sort

of police behavior that activists sometimes seek to provoke, knowing that insofar as it comes into view, it will be publicly condemned. A great deal of it never does come into view: that, at least, is what we suspect, and it is what makes the police so fearsome. Mere law enforcement is not fearsome, and though revolutionaries may well object to it, they are unlikely to win much sympathy. The far Left lives symbiotically with the "pigs," counting on each outburst of their brutality and then of our indignation.

But if official violence often generates spasms of national anger and self-reproach and "radicalizes" (a word that grossly exaggerates what actually happens) hundreds or even thousands of people, there is no doubt that over time it has a deterrent effect. Unless the political base of the Left (near or far) is very strong, police brutality will erode it, for ordinary men and women, part-time activists and camp followers, will withdraw and seek shelter. The aftermath of Chicago 1968 was not mass movement, but silence, disillusion, collapse. And it would be foolishly optimistic to imagine that large-scale student demonstrations are going to be easy to organize in Ohio after Kent State 1970. The story may be different in colonial situations, where the mass of the population is already alienated from the authorities and where a particularly brutal incident may trigger a desperate resistance. There, people are not shocked but mobilized. By contrast, the failure of mobilization in this country suggests how far we are (even in the ghettos) from a colonial situation. Confrontation politics, *especially when the police join in,* makes every sort of Left politics more difficult. That is a very important reason for demanding a police force that won't join in.

II

It is also a reason for condemning leftist provocations—like the recent mad campaign to "bring the war home." That effort seems already to have collapsed, though it may continue to have an aftermath of casualties. The career of domestic warfare has been brief and its various moments were so quick that they can hardly be said to constitute strategic choices in need of careful criticism. Yet it's worth setting out the record of those moments, even as it appears to an outside observer. I don't doubt that there is an inside story, too, more interesting than any I can tell. But it is important here to establish a certain category of militant acts: *those that one wants the police to deal with*. They must deal with them, as I have already argued, without "incidental" violence, without punishing the militants before they are brought to trial, without using far Left madness as a way of attacking other sorts of dissident politics; but deal they must, and, I am afraid, with our support.

The first moment was that of street confrontations, direct attacks (though not with lethal weapons) on massed police. These were intended, if that's not too strong a word, to open the way to full-scale white rioting, comparable to that of the blacks in the ghetto. Working-class youths were supposed to join in, drawn by the *machismo* of the militants and their manifest readiness to break away from the fears and inhibitions of their own past. None did join, so far as I know, during the first dramatic incidents. Indeed, nothing at all happened, except that the police, probably quite happily, fought back, and considerable numbers of young militants from the middle and upper classes were injured and arrested.

The second moment was that of collegiate "trashing." Now the ghetto imagery was acted out, not in any of our major cities, but in university towns like Cambridge and Madison. And now young people from the local high schools, some of whom must have been working-class, joined the militants, without any discernible political (there may have been social) motives, in breaking the windows and looting the stores of petty-bourgeois merchants. Another example of class war, and again unrelated to any political struggle.

The third moment brought urban terrorism, a surrender of any hope of mass participation, a search for efficacy in the Act rather than the Movement. Or perhaps it is not a search for efficacy at all, but only an expression of hysteria, fury, and failure. Thus far, the attacks have been aimed at buildings rather than at people. I'm not sure if that reflects some murky residue of moral principle or a peculiar sociological sophistication. American society, it may be thought, is built of steel and concrete rather than of men and women. That's wrong, but it's not an error we should seek to correct.

Now toward activities of these sorts, it's obviously not possible to engage in comradely criticism. Terrorists end without comrades. But before that, confrontation and trashing already revealed the collapse of any sort of collective discipline on the Left. I don't mean only (though this is important enough) that there is no mass party from which we can expel idiots, adventurers, and plain criminals. There is also no common forum within which strategic options can be debated, no "public opinion" that can exercise any restraint over individual enthusiasts, no stable body of leaders who can be held responsible for whatever happens. In the last years of the sixties, *laissez faire* triumphed on the Left. And just as in the economy, so in the political arena, and even among radical entrepreneurs, the triumph of *laissez faire* rapidly creates conditions that require state intervention. It also creates conditions in which the state—in this case, the police—can be sure of widespread popular support when they do intervene and may have such support even if they intervene inefficiently and brutally.

What happens to the militants then? With the beginning of terrorist activities, they went underground, and there, wherever it is, they lead a life that I can only speculate about. I find it hard to credit reports that there is an underground *movement*. There clearly are a surprising number of young Americans *hiding*, and there is undoubtedly a network of hiding places and a communications system of some rough sort for sharing information about how to do it. But the existence of a movement implies what is not at all apparent today, a planned and coordinated pattern of activities. Probably no such pattern can be developed without considerable above-ground support, and that the terrorists don't have.

They have, however, found a certain amount of sympathy within that curious American phenomenon, the above-ground underground, the highly publicized, perfectly visible subterranean world of the cultural revolution. Despite all the talk about repression, this largely legal "underground" life, typified by the "underground" press sold freely on Cambridge street corners and, I expect, in other cities, has been expanding rapidly. In a more accurate metaphor: the margins of American society are wider than ever before; it has never been easier (though a prolonged recession would change this quickly) to pursue a marginal career. A certain sort of occasional politics is one such career, and terrorism in its turn is marginal to that and might readily be assimilated or reassimilated to it were the terrorists ever to decide that bombing is not the true and only way. Some of them seem already to have made that decision. It is hard to see any other future for them: the cultural revolution is the only revolution they can hope to join.

III

The far Left has split so often, the tale of its internal wars is so tangled and so sad, that the rest of us can be forgiven if we allow ourselves to lose track of its various tendencies and their various ideological excuses. But one division seems to me important. Over and over again the line has been drawn (not always with the same people on the same sides) between those who are willing and those who are not willing to tolerate or encourage cultural marginality. The confrontationists, trashers, and terrorists have been among the tolerant, as well they might be. On the other side are all those groups that have tended toward one or another sort of Marxist (Trotskyist, Maoist) sectarianism, a tendency never free from puritanical impulses, but committed also to relate somehow to the "masses"—even to middle Americans. The sectaries have their own view of violence, which involves theoretical approval of it at some revolutionary moment sure to come but relatively far off and, at the same time, a refusal (mostly) to act violently here and now. The refusal, at least, brings them nearer to the near Left, and it calls to mind a political position on violence that I will try to restate briefly below. The position is worth restating not, I think, because of the Weathermen or the Panthers, whose future is dim, but because of the cultural radicalism with which we are likely to live for some time.

The new culture may indeed turn out to be as nonviolent as some of its friends and publicists suggest. But it is not now and never has been even remotely supportive of a nonviolent politics, or indeed of any politics that requires organization and long-term planning. The radicalism of "Do it!" has no room for prohibitions on trashing or terrorism or anything else. It is, instead, the mirror image of political (and why not of economic?) *laissez faire*. Against the background of extravagance and fantasy that it provides, there is something attractive about the young sectaries, unlovely as they may be in other respects, who recognize that there are some things that ought not to be done—today, anyway.

Judging from the rhetoric (and the tone, style, voice level, etc.) of their spokesmen, their press and movies, the fantasies of many of the cultural radicals are rich in violence. Though they themselves most often appear as victims, part of the fascination is with the possibility of striking back. Not, I think, with the possibility of winning some specific struggle against oppression—the radicals may or may not be victims, but they are not oppressed—nor with the possibility of joining the ranks of the oppressors, temporarily, as has been done so often, for a "good cause." There was a kind of stolid commitment to brutality among Stalinists in the 1930s, but nothing like that exists in the varied ranks of the cultural revolution. Theirs is not the politics of

the necessary murder . . . the flat ephemeral pamphlet, the boring meeting. . . .[1]

so much as it is the cult of experience, with violence, for some of them, as the penultimate experience. *Penultimate,* because attacks on the police, for example, are often said to have the happy effect of liberating the attackers, and there must be some further reason for doing that.

But perhaps freedom from moral inhibition is an end in itself: one acts violently in order to free oneself to act violently. It may even be thought that revolutionaries are created through such a process, but I know of no evidence for that claim. The end-state is some special sort of grace (or spiritual pride) rather than political effectiveness. Religious zealots have often in the past set themselves such a goal, and I expect it is even possible to reach it, though there appears to have been some escalation in the necessary means. In the seventeenth century, it was enough, according to certain antinomian sects, to fornicate, or at most to fornicate repeatedly, and so liberation was available to the millions. Freedom through violence is not so egalitarian a doctrine. It is, however, easily cheapened: violence is an experience that can be enjoyed and is most often enjoyed by its advocates at a distance and vicariously. Nor do I know many people ready to pay the price of a more serious commitment. The greatest violence of the cultural radicals is expressive, not practical. And with *that* the rest of us can live. The problem with the re-

---

[1] The phrases are from Auden's "Spain." George Orwell has some fine things to say about them in his essay "Inside the Whale," in *The Collected Essays, Journalism, and Letters,* ed. Sonia Orwell and Ian Angus (New York, 1968), vol. 1, p. 516.

vived and already corrupted cult of experience is not that it is likely by itself to produce more violence than we can endure or the police can handle, but rather that it provides no support for everyday politics. Instead, it sets the stage for once-a-month or once-a-year explosions, bursts of excitement shared through the media, with effects that no one even pretends to think about.

As putative victims, the cultural radicals are more appealing but no more helpful. They have produced a counter-image to the violence of our time, not pacifist but pastoral: the picture of a "green America," themselves the rightful inhabitants, beset by technologically advanced barbarians. As an art form the pastoral has hitherto been a specifically upper-class diversion, and it remains to be seen whether they can make it into anything else. Here I want only to note how innocent their picture of the future is—vaguely anarchist, sternly ecological, sexually bountiful, depopulated—and how incredibly far those who hold it have to go before they can even think of acting it out. Every practical decision lies ahead of them; every political temptation is yet to be encountered; every risk yet to be taken. Now the cultural radicals simply turn away from the violence of the American state and its enemies. If they have to, will they be able to criticize, organize, resist, and still maintain their peculiarly peaceful vision?

IV

Politics, once again, is the art of minimizing violence; a man begins to be a politician as soon as he tries to get what he wants without hitting somebody. But social change, we are told, requires that somebody be hit; nothing has ever been achieved . . . etc. There is some truth here, but questions need to be asked in each case—just as questions need to be asked of the police. At what point, in what (precise) circumstances was violence used? Who used it first? How many people were injured? With what (concrete) results?

A disciplined mass movement is much like a strong and legitimate state: it uses violence artfully and economically; often the mere threat is enough; "incidental" brutality is barred or punished; innocent people are not attacked. The classic vision of the near Left is of such a movement, growing steadily over time, winning massive support by means almost entirely peaceful, but then *because of its strength* being challenged by the authorities and forced into strikes, demonstrations, perhaps even street fighting, and so on. It was never thought that this challenge could be successfully met, however, unless substantial sympathy had already been won among the police and within the ranks of the army. Even "radicalized" citizens do not conquer policemen and soldiers; they win them over (or they don't). Hence violence is but one instrument of politics, and by no means so ready a choice for the Left as for the authorities.

But the movement grows slowly, and so the short cut is the first and the perennial political temptation. Violence always looks like the easiest short cut. Sometimes this or that group of militants hopes to seize power all by themselves without waiting for anyone else, or to terrorize the country and win major concessions on behalf of the others; or they hope to provoke police repression and so goad their potential but passive followers into revolutionary action. The last of these is the strategy most often attributed to contemporary militants. They intend to expose the "true character" of American society, it is said, to drive the authorities into an openly fascist posture, and so force the rest of us to resist. What is there to say? I doubt, again, that this is a serious intention, though it is a strategy that might conceivably work if (1) the true character of the United States government were fascist, and (2) the mass of the American people were already profoundly alienated from it. Neither of these conditions holds. Nor do strong and legitimate conservative regimes turn fascist in response to leftist provocation; I know of no historical cases. And if they did, that would surely be a disaster and not a success for the Left. No fascist regime has yet been overthrown by rapid revolutionary mobilization; nor is there time for long-term political organizing; in the short run the militants would be dead. It is, all in all, a policy of insanity compounded.

I have already described the actual results of this "strategy" in the United States today. But there is more to say about it, a classic leftist position to reiterate against it. However easy it may be to throw a brick, use a gun, plant a bomb, and however much the cultural atmosphere encourages—and publicists, editors, screenwriters, etc., applaud or "understand"—such activities, they are not likely to be engaged in by large numbers of people. They don't build a movement, because they don't commit men and women to daily work for the cause. Terrorism is not and cannot be a collective effort. It enhances the self-regard of a militant elite, and then the members of this elite celebrate their psychic or moral liberation. But the rest of us have no reason to expect that if they were ever to seize power, they would be any different than the people they call "pigs." The one group is as unlikely as the other to liberate *us*. That we can only do for ourselves, through organizations in which we are actively involved and whose policies we share in making. "The liberation of the workers can come only through the workers themselves."

The liberation of the militants is a waste of violence, just as the beating up of the militants by the police is a waste of violence. Neither is a victory for the rest of us. The only difference between the two is that in the second case there are review boards and presidential commissions to condemn the waste, and in the first there are none. The "movement"—for the moment—offers us no recourse. The democratic state does. So *laissez faire* on the Left makes statists of us all, and our most immediate duty is to be honest about where we are. Inevitably, the police will impose the discipline we have failed to build: the problem of law enforcement—not a problem we are at home with—has been moved onto the political agenda of the seventies. Significant social and political change may well await its working out—may well await, that is, the restoration of a framework of civility within which the police, the militants, and the rest of us can live, more or less, together.

(1971)

9

Notes for
Whoever's Left

I

Perhaps it is presumptuous for those of us who were often sharply critical of the New Left to address ourselves now to its scattered and disorganized followers. Yet among them are future friends and allies—who have begun to make criticisms of their own—and it is important that we find a way to talk together, sooner rather than later. The last years of sectarian in-fighting, ideological debauch, and pseudorevolutionary violence have taken a heavy toll, not only among New Left militants. There are signs today of a massive withdrawal from political involvement, a spreading mood of cynicism about the possibilities of political success. It is clear that any sort of sustained leftist activity is going to be extremely difficult. *In the New Left fall/We suffered all.*

"The political scene has seldom looked more dreary"—so Christopher Lasch wrote last October in a *New York Review of Books* piece, "Can the Left Rise Again?"[1] His is probably the

[1] Vol. 17, no. 6 (October 21, 1971).

best statement on the present condition of the American Left, and I shall use it as a guide, even while marking off some areas of disagreement. Half a year after he wrote, things look more dreary still. There has been no significant activity at all, nothing but a few ritual efforts by leftover cadres unable to command any serious response from the ranks. One has the uneasy sense that as an organized force the New Left of the sixties has simply disappeared, leaving behind a small number of embittered terrorists and a few isolated and impotent sects.

That is certainly wrong, for leftish opinions of a more or less "new" sort are widely believed today, and even fashionable. I still meet people of all kinds (and in surprising places) whose views are "far to the left" of my own. But I don't meet them, nor does anyone else, at meetings or demonstrations. It's probably not unfair to say of most of them that their opinions are all that is left of their leftism. So far as organization and activity go, they have been radically disillusioned by the ugliness of the last few years, chastened by failure—even when it wasn't their own failure. They have lost any sense of what ought to be done and any willingness to do it. Well, so have we all been chastened, and no one at this moment has a clear program. When the revolution doesn't happen, it's not easy to find one's way back to everyday politics. Some people—leftist dropouts—simply won't find their way. And the result is a downward spiral of frustration and retreat that gets wider and faster because re-entry into the mainstream of American life, or into its cultural margins, is much easier than a leftist reorganization. A few radicals are left stranded on peaks of militancy, easy victims for the authorities, while life on the plain returns all too quickly to "normality."

Again, that is not an accurate picture: American life has changed a great deal during the past decade, and much of the change is the achievement of the Left, in some broad sense of that term. What is crucial and disastrous, however, is that the Left has had no *victories*. A victory is an achievement that gets credited to somebody, that some movement or party claims as its own, on the basis of which it advertises itself, recruits new members, even seeks a share of political power. In politics, nothing is more important than winning victories, and the Left has not won any. Yet look what has happened: the country has been turned around on the Vietnam war (though the war has not yet been ended); the position of black people in America has been radically transformed (though we are still far from racial justice). In both cases, the peace movement and the civil rights movement had a share in doing what got done. In both cases, these movements played a leading part in the political mobilization of masses of men and women. But these movements spurned the (admittedly incomplete) victories that could have been had, and as a result they were not able to establish an independent or stable presence in American life.

Here I want to suggest a disagreement with Lasch, which may have implications for the future. He argues that the turn toward a more militant politics in the middle and late sixties occurred because of the failure of "polite reformism"—"the failure of the peace movement to end the cold war or to prevent Vietnam, the failure of the civil rights movement to achieve racial justice, the failure of reform Democrats to reform the Democratic Party." It seems to me that in at least two of these areas—the war and the fight for the Democratic party (I'll have something to say about civil rights later)—the critical struggles came *after* the New Left had turned to an ultra-radicalism that made its effective participation dubious or impossible. It's not that "we" (the "polite reformers") failed and "they" became radical, but that "they" became radical and then "we" failed. On the other hand, it has to be admitted that such liberal-left efforts as the McCarthy campaign were organized at least partly in response to pressure from the far Left (though also in response to the widespread demand for an alternative to the compulsions of far Left politics). Nor do I want to underestimate the sense of urgency and despair that drove thousands of young Americans steadily leftward. Their politics was not merely willful, though one wishes they had sensed the limits beyond which it became dangerous (to others as well as to themselves).

Throughout the sixties, as Lasch admits, there existed a symbiotic relation between liberals and radicals and between near and far leftists. We rose and fell together—1968 was a crucial year for all of us—and there are lessons in that fact that I will try to draw out later on. For the moment, we all must live with our failure to understand and profit from those relations and to take what might have been had. And we must be frank about the extent of the failure, for it has to do not only with what happened or didn't happen in the sixties, but with our future prospects. Ten years of militancy and upheaval have left virtually no institutional residue: no ongoing organizations, no coherent or nationally recognized cadres of leaders and spokesmen, no respected New Left journals, no public network to keep radicals in touch, no established or prototypical way to be *and keep on being* a radical.

Against this, it might be said that the counterculture represents the real success of the New Left—and it requires nothing more from the liberal world than tolerance. What is involved, or so we have been told, is an irreversible transformation in the consciousness of the young. If that were true, the New Left would be a force to reckon with, and in the very near future. But it is far from true. Whatever value and impetus the counterculture once had derived from the political movements with which, however tenuously, it was associated. Deprived of that connection, it is literally valueless, with no substantive political or moral content, with decreasing power to attract the young, and readily susceptible, as we have begun to see, to an almost infinite debasement. "Desperate and bitter, brutalized by drugs and police," writes Lasch, "the youth culture sinks into the underworld and [its participants] become increasingly indistinguishable from the lumpen-proletariat." A harsh judgment, but probably accurate. Nor are other forms of fashionable leftism likely to prove any more sustaining. A powerful political movement can create a culture-on-the-side, where distinct styles of writing, singing, dressing, and above all of talking are cultivated—as the Communist party did during the days of the Popular Front. Such a movement may even feed on the participants in this culture. But it cannot live on them. There is no substitute for the commitment and discipline bred by significant action.

II

What made the sixties exciting, what made its (limited) achievements possible, was the fact that different groups on the Left shared a sense of what needed to be done. Issues that had been avoided for a long time suddenly seemed unavoidable, and despite important tactical and strategic disagreements, a number of specific activities won large-scale support. Efforts to explain these activities in ideological terms or to fit them into a world-historical framework, however, did not work: no position won wide acceptance or proved even minimally persuasive. Not that people didn't try to persuade or to be persuaded. But the political forces set in motion in the sixties were extraordinarily diverse, with different histories, class compositions, political styles and goals, and it was never likely that a single or unified explanation would account for all of them or establish guidelines for their cooperation. Today, the common commitment to issues also has disappeared. We live with a new uncertainty, intellectually refreshing after all the "correct ideological positions" that have been inflicted upon us, but also politically enervating. I want to try to suggest some of the reasons for the uncertainty and some ways of living with it, without pretending that my suggestions have any authoritative basis in personal or collective experience. The near Left has not triumphed either; nor are any of its leaders writing manifestos just now. It would be a great mistake, nevertheless, not to try to draw lessons from the recent past or to specify, however tentatively, some ongoing difficulties.

1. At the center of uncertainty is the problem of relations

with the liberal world. I don't mean, the liberal establishment, whatever that is, but more simply, liberal politicians and their present constituencies, from whom leftists have grown increasingly estranged (not without reason). Historically, the Left has stood in a complicated double relation to the liberal world. To put it crudely, we have said, first, that we want something more than and, second, that we want something differences. The larger society, if it is ever to come into existence, still requires an integrated movement. As for the smaller thirties and again in the mid-sixties, significant social reform has come when there was organized pressure from the Left—or at least, pressure (from workers and blacks) in the organization of which leftists played a major part. The seventies will not be different; liberalism is, if anything, less self-sufficient than ever before. It will be one of the tasks of militants trained during the past decade to build pressure on liberal politicians in the name of immediate and limited programs. I have in mind the sort of thing suggested in a recent essay by Leon Shull and Stina Santiestevan appropriately entitled "What Do We Want Right Now?"[2] The trouble is that many liberals don't want anything right now, except peace and quiet. We suffer from their weakness, as they will suffer from ours.

The second demand—for policies different from those that liberals readily choose—has to do with the identity of the Left itself. I will have more to say about that later; now I only want to point out how the search for an independent identity has time and again led leftist parties and sects to turn liberals into their "chief enemy," and to deny that there is any continuity at all between liberal success and radical aspiration. That there are important discontinuities between the two cannot be doubted. Nevertheless, liberals and leftists appeal to the same groups of people and share many of the same concerns—for liberty, above all. The two groups can go and have to go a long way together, whatever the tensions that arise in the course of their cooperation. Under present and foreseeable

2 *The Seventies: Problems and Proposals,* ed. Irving Howe and Michael Harrington (New York, 1972), pp. 468–486.

conditions, it just can't be the case that liberals are the chief enemies of the Left. If we are vital to their success, they are vital to ours. The liberal world is still the only possible base for leftist activity. And the only organizations through which we can hope to win victories are liberal-left coalitions, such as the civil rights and peace movements were before they fell apart. What happened after they fell apart suggests dramatically enough that we are not ready for the "pleasures" of going it alone.

2. The most painful moment in the history of the sixties came when white liberals and radicals were expelled from the civil rights movement. It's not difficult to understand the motives of the black militants who seized control at that time and turned the movement into something very different from what it had been. Nor is it difficult to sympathize with some of those motives. The immediate result, however, was disastrous for the Left—not only politically, but intellectually and morally. The last few years have seen a succession of racial and then ethnic self-assertions with which radicals have been utterly unable to come to grips in any serious way. The prevailing tendency (among many liberals too) has been to grant a kind of *carte blanche* to any "oppressed" group whose militants adopt a radical rhetoric, as if there were no principles by which their particular demands might be judged. Indeed, there are principles having to do with equality, mutuality, and political responsibility, and radicals should never have hesitated to apply them. But the principles that relate specifically to racial cooperation are newly in doubt. Until the mid-sixties the Left was, at least programmatically, an integrated movement, and its goal was an integrated society. There were few things about which we were more certain than those two; we readily made judgments in the terms they set. Today the Left, or whatever remains of it, is deeply divided along racial and ethnic lines, and the desirable future balance of integration and separatism is painfully unclear, at least to me. No one that I know of has a compelling vision.

The Left is face-to-face with pluralism—a notion savagely

repudiated by many leftists only a few years ago. But when we seriously begin to discuss its possible forms, we shall have to admit, I think, that the complete abandonment of integration by many black militants, and then by whites in their tow, was terribly wrong-headed. There has to be some balance between the larger society in whose running we must all share equally and those smaller societies where we cultivate our differences. The larger society, if it is ever to come into existence, still requires an integrated movement. As for the smaller societies, the Left has not yet faced the fact that there will be many of them. It's not only a question of blacks and Chicanos, but of Polish, Irish, and Chinese Americans, Jews, perhaps even WASPs—all caught up in a small-group nationalism, sometimes attractive, sometimes not, toward which the Left has no comprehensible position. In trying to understand what is going on, we might begin by giving up the word "racist"— at least until we have thought again about what its opposite is and what we are.

3. Participatory democracy has not, to put it mildly, lived up to the expectations raised by its spokesmen. On this point, Lasch writes with especial force, and I have little to add to what he says. The question, again, is one of balance: between the need for local autonomy and participation and the need for unity and leadership at the center, where local members can't be present and have to be represented. Obviously, a balance of this sort existed even in SDS, which made such a fetish of primitive democracy, but few would say today that it was the right balance. Decisions of enormous importance were in fact taken on the initiative of relatively small groups of leaders, but not in such a way that these leaders were easily identified or, more important, held responsible for what they did. Partly for this reason, the organizational structures of the New Left were always shapeless and weak, until the sects, in response to this weakness, reverted decisively and foolishly to the Bolshevik style of democratic centralism. But sectarian discipline is as inappropriate to any political movement that hopes to be

democratic as is New Left amorphousness. We need a new organizational strategy (or strategies), and this is not likely to be developed in isolation from some sort of renewed activity. I would argue only for a representative and responsible leadership, whatever the precise structures over which it presides. For the moment, of course, leadership is not the problem. Organizational needs are very limited: an open forum, above all, a place to regroup.

4. In the sixties the Left experimented with a variety of constituencies: the black and white poor, the working class, the "new" working class, middle-class students, the middle class itself. What were the results? I don't want to sound like the old man who commented bitterly about sex: "It's been tried with a man and a woman, and a man and a man, and a woman and a woman, and a man and two women, and a woman and two men—and it's never quite satisfactory." But the truth is that none of these constituencies was by itself quite satisfactory. Though each new social base was justified in elaborate ideological terms, none of them turned out to be the true and only social base. And today the Left seems baseless, without established roots in any particular community. Across the country, there is no clear social direction.

Nor are there likely to be, it seems to me, in the near future. Left politics is certain to survive, but on a fairly small scale, and we must seek support where we can find it, *wherever* we can find it, without ideological presuppositions. My own sense, as I have already indicated, is that the place the Left is most likely to find support is among liberals—old-time civil rights and peace activists, mostly middle-class, and middle- and lower-level people in trade union and black organizations. That means—where it has always been found. But it is also certain that much support must continue to come from students, even if the immediate impact of political disillusion is most sharply felt on the campus (where political fantasy was most free and extravagant a few years ago). And we may be able to make peace with at least some of the ethnic militants, though not on just any terms. Before that, we need to make a little peace among ourselves. Tolerance and openness are the key.

III

I have waited until the last to say anything about socialism. For me, and for many of those radicals who continue to think of themselves as political activists, socialism remains the directing vision of the future. And it is still our hope that the modest projects presently available to us can somehow move us closer to that goal. Lasch argues forcefully that the special effort of the Left *right now* should be to make contact with people where they work and to relate our politics to their work (and to our own). He is elaborating on an old socialist theme and describing what would certainly be a central feature of the activity of an autonomous socialist movement. To bring together politics and work, that is, to set a democratically run economy alongside a democratically run state —here is the core of socialist aspiration. It is important to remind ourselves of that, and to remind others.

But socialism is not, unhappily, an immediate political issue in the United States today. Lasch calls for a "fusion of community politics and trade union politics" as the only way of making it an issue in the near future. That is an attractive idea, but I am hard pressed to understand what sorts of activity may lead to or represent such a fusion. Lasch's argument that it is already represented by the student movement, since students are organized where they work, seems to me radically unconvincing. In the first place, students don't work, that is, don't earn a living, which makes a great deal of difference. The discipline of the university is not at all like that of the economy: students are not an oppressed class (few of them could ever have believed that, nor does Lasch), and if many of them are apprentice technicians and professionals, it has to be said that their apprenticeship is more like a moratorium on than an initiation into routine economic activity. And second, students were organized successfully only against the war. Syndicalist efforts to win control of the university "workplace" ran a very poor second. Properly understood and constrained, such efforts might have been politically valuable; as it was, they did not meet the requirements of Lasch's "fusion." Right now, "self-management," the socialism of adults, is an end without a visible means. Our most important work is still theoretical.

Our practical work will focus differently, for we are in no position to choose our causes with reference only to our ideology. We shall be involved in difficult battles over civil liberties, unionization, civil rights, social welfare, ethnic pluralism, crime, housing, education—and, endlessly, foreign policy. In all these battles, we can bring socialist perspectives to bear; we have an ethic, a literature, a history to draw upon. But we have no easy answers to the hard choices that we, and others, will face. Let me give just one example: when Chinese American parents in San Francisco oppose school busing in the name of the cultural autonomy of their community, what is the right socialist or radical response? I know where we stand on the question of who should run the factories, but here I am uncertain. Nor do I see any easy move from one choice or the other to a socialist politics. We still have to look forward to the time when our vision actually integrates our day-to-day activity. Right now, it only draws us together to talk about and plan for that time.

(1972)

10

The Peace Movement
in Retrospect

Our politics has been marked by a strange logic, a flawed syllogism: the people (a clear majority of them) supported the government; the government carried on the war; but the people did not support the carrying on of the war. At the same time, they did not support the peace movement either, at least not in sufficient numbers. At crucial moments—after the Tet offensive or during the Cambodian invasion, for example—the movement undoubtedly had an impact, but it did not have a decisive impact. The country turned its back on the war, without ever giving those who opposed the war a chance to end it. Nixon finally made the peace and reminded us (and the Vietnamese) with his last savage burst of bombing that he can unmake it too.

But perhaps all this puts the emphasis in the wrong place. When in our history have so many citizens challenged so central a government policy? When have so many fled the country, deserted the army, disobeyed the law, organized, demonstrated, agitated—refused to be "good Americans"?

When before this have veterans marched against a war in which they had only just stopped fighting? For all the difficulties and tragedies of American politics these past ten years, there is something enormously heartening in the way a part of the country, at least, rallied against the Vietnam intervention. *But why didn't the peace movement win?*

For purposes of analysis, the movement can be divided into three parts. First, there were the students. They provided the crucial manpower; their energy, commitment, perhaps above all, their free time, made both mass demonstrations and presidential campaigns politically possible. Their characteristic activities, however, took place on their own campuses. It was there, appropriately, that the educational work of the movement began—with the teach-ins of 1965—long before ordinary Americans had become experts on Vietnam. But the university was more than an intellectual home. It provided the only political space that opponents of the war could successfully contest, and many of the most dramatic contests necessarily took place there. Given so precise and peculiar a location, it was inevitable that the immediate issues would be local in character, related only indirectly, if at all, to the larger purpose. Student protests focused on university cooperation with the draft, campus recruiting for the armed forces, ROTC, investment policy—and at the same time, black studies, open admissions, community relations, and (always) amnesty for the demonstrators. In the sixties it was terribly hard to concentrate. But it was always the war that fueled the protest. And it was the growing horror of the war, the daily humiliation of reading the newspaper, and the ineffectiveness of campus action (for whatever student radicals said about university complicity, it was clear that professors and administrators were not determining the nation's policy in Vietnam) that drove some students to an increasingly rancorous, intolerant, and violent politics. It is difficult even today to measure the effects of that politics. No doubt it mobilized many worried adults to seek alternative ways of opposing the war, "to bring their children home" before a generation ran amuck. No doubt also it frightened and antagonized others, especially those whose children were not involved, whose children were more likely to be drafted than to go to college. Student politics clearly revealed the class base of the peace movement —which turned out to be one of the most important things about it.

The second part of the movement was the creation of the Old Left. A whole series of operations was staffed and led by the members of established liberal-left and radical political or religious groups: the National Mobilization, Vietnam Summer (1967), the Moratorium (1969), Clergymen Concerned, Women's Strike for Peace, and so on. Focused more narrowly on the war and, whatever their internal sectarianism (of which the less said the better), tending toward a pragmatic and popular politics, these groups carried the antiwar movement off the campuses and reached out to a larger middle-class constituency. Veterans of the Progressive party and the Popular Front, socialists, Trotskyites, members of SANE, ADAers, Quakers, students radical and liberal, moderate and militant, Eastern (and Western) intellectuals: all these they brought together in massive reunions in the streets of Washington, New York, and San Francisco. They also organized in their home communities, fought local referendum campaigns, raised money for newspaper ads, signed their names. Theirs was a politics of demonstration. They sought to create and then to express antiwar sentiment, leaving it to conventional politicians to respond to these expressions, to seek the support of the movement, by directly challenging or changing government policy. Politicians did respond, not very bravely, with a few exceptions, but in ways that opened new opportunities for political action. The Old Left provided a bridge across the chasm that usually separates radical agitation from party politics, and its younger and more moderate followers marched eagerly across.

Some of the leaders (and the student militants in SDS) held

back, worried that the purity of the movement and its radical ideology were being compromised. But in fact, groups like the Mobilization were not ideological or particularly radical—even if there were speeches about American imperialism at every rally. Their real task was to generate a mood of hostility to the war or at least of doubt and anxiety about it. That's why it was so important that the crowds be large, look respectable, and so on; the speeches hardly mattered. And eventually those multitudes of respectable people had to find a way into the political system.

It was the third part of the peace movement—the presidential campaigns of McCarthy and McGovern—that brought it closest to success. For the American Left, conventional politics is itself an achievement. And what was called the "new politics" was entirely conventional: a matter of precinct organization, hard work on city and county committees, endless canvassing and phoning. It seemed new because the people were new and were not defending their interests so much as advancing a cause. They were not pros, though no doubt some of them will become pros. In any case, their first efforts were exhilarating. Here was a chance to fight for radical goals within one of the major parties and to hope to win. Unhappily, the movement won only the immediate victory and not the final one. In 1972, building on the achievement of the McCarthy campaign, its partisans seized the Democratic party. But that extraordinary victory split the party and revealed the precise limits of movement strength.

How much support did the movement have? From the local referenda of 1967 (in San Francisco, Madison, Cambridge) to the McGovern campaign of 1972, and in the opinion polls throughout those years, the figure hovered just under 40 percent. There seemed to be a barrier fixed to its growth, even while the activities of the movement expanded enormously in scope and antiwar agitation was moved from the margins to the very center of our national life. And that barrier had a recognizable social and political character: the movement was unable to penetrate significantly into the ranks of organized labor and of those working-class ethnic groups that have traditionally been the strength of the Democratic party. There are so many reasons for this that it is probably right to say that the failure was over-determined. But it needs to be talked about anyway.

The most immediate reason is also the most obvious. The Democratic politicians most highly regarded by American workers were themselves implicated in the Vietnam war and unable or unwilling to break away from it. For years this was a Democratic war, and a whole generation of liberal leaders (whose domestic achievements had earned them that title) was discredited and destroyed by it. Trapped by their past commitments or still believing in the Cold War ideology they had so long espoused, they did not repudiate the horrors of Vietnam even after Nixon had made those horrors his own. They hoped desperately that the fighting would end before 1972; they twisted and turned to get off the hook; they never made a moral decision against the war (although they stopped defending it), and they turned out to have made the worst of all possible political choices. Had they behaved differently, a unified Democratic party would certainly have triumphed in 1972. Instead, they pushed their constituents toward Richard Nixon.

But their constituents probably did not have to be pushed very hard, for the peace movement antagonized many Americans whose support it could not do without. Partly this was because of what movement partisans did: the continual provocations of self-righteousness and zeal, the senseless challenges to working-class patriotism, the deliberate linking of antiwar and countercultural politics. Partly it was because of who the movement partisans were. Opposition to the president on foreign policy traditionally draws on a very limited constituency, an elite of the concerned, inevitably self-chosen, largely from the upper and middle classes. The peace movement was the product of a class society. Students and intellectuals, suburban men and women, amateurs with time and money to spend: it was not only their creed that was offensive to American workers but their persons. Behind George Meany's pique there was a powerful class hostility, which he and others were able to exploit. Important labor leaders and many thousands of workers supported the peace movement, especially in the unions with a strong leftist heritage (like the UAW), but they never made it their own. Its tone was always middle-class, academic, suburban. Only in the army itself and among veterans, exiles, and deserters did the movement seem classless. But there it was also very small.

The crucial battle was fought for the hearts and minds of American liberals. Its long-term result is still unclear. It may give rise eventually to a politics more humane and more *worried* than the complacent and technological liberalism of a decade ago. Immediately, however, the divisions in the liberal world made it certain that the war would go on, and the ongoing war further deepened the divisions. Nixon was the sole beneficiary of this process. Many people said that he had to end the war in order to win re-election. But in fact it was his refusal to end the war for four long years that guaranteed his electoral triumph.

Except for the reformed Democratic party, the peace movement has left behind no institutional residue. Those of its activists who thought they were building a political base for the future will probably be proved wrong. The "new politics" will survive, but only within the Democratic party and as a part of one or another future coalition. What has now been demonstrated is that it cannot win by itself and that it cannot grow by simple expansion. But that is only a practical lesson. The real effects of the struggle against the Vietnam war are moral. Those effects are what is usually being referred to when people talk of a crisis of authority in this country. That means: the American government is today less believable and less "obeyable" than it has ever been. That is the price its leaders have paid for their war. We have to be glad of that, for what sort of a country would America be if they had had to pay no price at all, if their lies had been believed and their crimes justified by *all* the American people? In this sense, the peace movement was a partial success. It made the waging of the war morally costly; it re-established the importance of moral and legal discourse about military conduct and political authority; it created a larger constituency of conscientious men and women than the country had seen before (and made some of them into experienced politicians); it began, perhaps, the long process of setting limits to what governments can do and to what citizens must bear.

(1973)

11

The Pastoral Retreat
of the New Left

I

What has happened to the young men and women of the New Left? The movement is invisible these days, a specter regularly invoked only in neoconservative writings. Where have all the "kids" gone? Many of them are simply burnt out, exhausted (long ago) by sectarian infighting, corrupted by revolutionary visions, lost in the rubble of the counterculture. It is an old American story. The flames of the political Left don't burn steadily in this country. They flare up; the young are consumed. And all that remains, afterwards, are the ashes of political withdrawal, cultic and sectarian conversions, and personal opportunism.

But that's not the whole story. Apocalypse never is the whole story. Even in capitalist America, there are ways of growing up on the Left; there is interesting work to do. In the universities, for example, where the New Left began, old New Leftists constitute today a significant intellectual current

among younger faculty members (with little influence, iron-
ically, among undergraduates). Many of them are trapped by
the academic depression. Some of them are trapped too, at
least for the moment, by the rather scholastic Marxism,
Francophile or Germanic, that they first discovered in their
flight from the Leninist orthodoxy of 1969 and 1970. But
theirs is a serious intellectual discipline; they are not burnt
out or mad in the streets; and one has only to read their
articles to realize that they stand at a great distance from the
politics of instant satisfaction. I am often amazed by the
strength of their professional commitments. This is a form of
adaptation and survival, however, that is available only to a
small number of activists in a movement that was not by any
means scholarly or intellectual in character.

There is a second pattern of survival that is more important,
I think, and more revealing of contemporary American life.
A large number of the activists of a decade ago have found
their way back into politics, or at least into a subpolitics of
small-scale worker and consumer organizing. They have moved
from *revolution now!* to good works, and this movement is
also a return, though everything is different now, to some of
the original projects of the New Left. Across the country there
exists today an extraordinary network of schools, day-care
centers, food cooperatives, tenants' unions, consumer organizations, neighborhood alliances, locally run newspapers, union
organizing committees (aimed mostly at service and clerical
workers), women's groups, and so on. Devoted to the politics
of everyday life, locally organized, many of these groups are
financed with government, foundation, or church money; they
employ VISTA volunteers; they have trade union connections;
they run successful campaigns; they have achieved, as the
New York Times recently reported, "a major realignment of
urban political forces." Taken together, the local groups represent a new kind of interest or pressure politics, or a radical
extension of the old kind, and some of them are already incorporated into the institutional structures of the welfare
state. Nevertheless, their leaders, their staffs, their organizers-

in-the-field are recruited, to a significant degree, from among the radicals of a decade ago, many of them from the once terrible SDS. They know one another, keep in touch, and are guided by an ideology that has a familiar ring to anyone old enough to remember the *early* 1960s. This is the New Left in pastoral retreat.

They are, as they ought to be, a chastened group. Locked into small and mostly precarious organizations, they have learned the art of making do, and they have acquired a sense of limits. Certainly, their current activities fail to realize what seemed to be the major opportunities of the sixties. Except for some union work, the new organizers don't reach into the black community at all. And theirs is not, by any means, a "poor people's movement." They have not done very well among welfare recipients, tenants of public housing projects, unemployed men and women. The groups they are able to form and sustain mostly involve (relatively small numbers of) better-off workers and members of the lower-to-middle middle class. And the politics of these groups is clearly reformist; the neighborhood alliances often take on a kind of "community uplift" character. Self-help against crime, the defense of old residential areas, improvement of local services, beautification; these are their goals, to which the organizers too must stand committed. In their late twenties or early thirties now, these middle-class militants have found a place for themselves closer to home than some of the places they experimented with in the early days of SDS community organizing.

They are not *narodniks* anymore, and yet the impulse is the same: to begin locally, to work in neighborhoods. By and large, the issues with which they deal are the same too. They are concerned with services and consumption—housing, utilities, recreation, safety, taxes—far more than with wages, working conditions, or work itself (though the forms of their politics easily reach to questions of safety at work or discrimination in hiring).

And their values are the same. The organizing strategies

taught in schools like the Midwest Academy, the principles formally endorsed or informally defended by groups like Massachusetts Fair Share and Arkansas ACORN are still those of participatory democracy and anticapitalism (the latter blurred these days into a kind of populist anticorporatism). The organizers are not interested in building constituencies of passive members; they still hope to teach the men and women they organize the skills necessary for a politics of self-assertion and the knowledge necessary for a radical understanding of social structure. This is their version—it is a protestant version —of the cure of souls. They dream of a priesthood of all believers; they are committed, still, to the abolition of their own ministerial roles. And yet the modern welfare state, like organized religion, seems to require that people be ministered to—organized, represented, defended—in the face of state and corporate power. The ministry of the Left is made necessary by the administration of the welfare system. I want to try to explain why that is so and how the roles into which these old/new activists have fallen are both very important and very constraining.

II

The peculiar difficulties of participatory democracy as an organizing strategy were much discussed in the sixties and are well known today, certainly to contemporary organizers. Their own commitment is made in the teeth of their knowledge, with considerable sophistication, and they meet the difficulties, I think, with more grace and less hypocrisy than their predecessors. I remember attending a meeting of the Newark Community Union with an SDS organizer in the mid-sixties. "Local people" presided; the organizer, resolutely self-effacing, placed himself in the back row; and the participants had to sit sideways on their chairs so that they could watch the reactions of the real chairman of the meeting. No doubt, the problems are the same today. Fewer local people come forward than the organizers hope for; few of them are willing to take on leadership positions; the leadership they provide is hesitant and uncertain; the organizers cannot go away.

Insofar as local people come to appreciate the uses of militancy and learn organizational skills (and there has been a lot of learning in the last decade), other sorts of problems arise. Members take a firmer hold, insist on a voice in shaping policy, and reveal in their political positions some of the immediate effects of inequality and its accompanying resentments. Some of the new consumer groups, for example, were steered away from Proposition 13 politics only by the most energetic interventions of their (leftist) staffs. There is, then, a certain tension between political education and participatory democracy. But the day-to-day difficulties of organizing have more to do with time and competence than with policy. This kind of politics, like every other, is a routine; its success depends upon one small victory after another—getting a new playground, forcing acceptance of an affirmative action program, stopping a hike in utility rates, winning a National Labor Review Board election in a hospital or office. These victories require professional and semiprofessional skills, and it is my sense, though I have only watched them from a distance, that contemporary organizers are, mostly, willing to display these skills (foregoing strict ideological rectitude) and to reap the organizational benefits that go with them.

Hence they drift, inexorably though always reluctantly, toward a kind of advocacy politics. They are more likely to identify and represent a constituency than to mobilize its members for active struggle. Or, they mobilize a relatively small group of men and women who then provide them with a kind of warrant to act on behalf of similar people in a particular neighborhood or city. And action on behalf of others, once revolutionary adventures have been forsworn, imposes a new pattern of political work. The organizers aspire to be representatives of the unrepresented, but they are also, inevitably, bureaucrats without offices. Though their political line is populist in character, focused on landlords, utility companies, and corporations, their practical activity quickly brings them into a complex adversarial relationship with public officials. Swinging freely at capitalism, they find themselves in a clinch with the welfare state.

What happens is simple enough. Social conflict in the United States today has few direct forms. Everywhere, it is mediated by the state. Municipal housing codes, rent control, affirmative action, NLRB regulations: all of these shape the course of particular struggles, and they shape it in ways that make advocacy and legal representation as important as political mobilization. Anyone trying to work with tenants, for example, quickly finds himself entangled in the rules that govern the repairs that landlords have to make and the rent that tenants can withhold. The rules are complex, and they require interpretation. Soon the organizer is negotiating with city officials; soon he has to bring in lawyers. It will sometimes be useful to organize a demonstration at a city council meeting. Just as often, it will be useful to go to court. And a single tenant withholding rent makes as good a court case as a thousand tenants withholding rent. Once the negotiations and the litigations begin, there is little, in any case, that the tenants can do for themselves. A few of them, like prison lawyers, become experts on the housing code. The rest sit and wait, while the organizers and the lawyers do their work. The longed-for confrontation between a citywide tenants' union and the big holding company that owns hundreds of apartment buildings never takes place.

This is a reiterated pattern. Even in union organizing where constituencies are clearly marked out and it is necessary ultimately to win a majority vote, the day-to-day activity of the organizers, and of their opponents too, is closely regulated, and crucial decisions are made by courts, commissions, and appeal boards. If old New Leftists turn up in law school, this is not necessarily because they are planning to sell out (at a high price), but because everywhere they turn politically, there is lawyer's work to do.

III

It is one of the purposes of the welfare state to transform struggles over group interests into litigation over individual entitlements. The transformation has worked fairly well in the past, in part because individual members of groups unable to defend their interests have also been unable to claim their entitlements. This is true both of groups without social or economic strength, like tenants or hospital workers or secretaries, and also of groups without clear socio-economic identities, like people driving a particular make of car or living along the banks of a polluted river or working in the service sector of the economy. The pressure on the welfare state today, about which neoconservatives complain so much, doesn't derive from any significant rise in the level of entitlement but rather from an increase in the range of men and women claiming what they are already entitled to. The organizers I have been describing, along with the lawyers who join in their work, are *agents of entitlement,* acting on behalf of the members of powerless or dispersed social groups.

I have called their relation to government officials "adversarial"; it is not in any simple sense oppositionist. They represent claims to which public officials are not yet responsive, but to which they are, in principle at least, committed to respond. And, after all, welfare bureaucrats do have an interest in the expansion and elaboration of the welfare system. So the organizers outside the government are likely to find collaborators inside. They win victories as often by negotiation as by political or legal confrontation: small victories, to be sure, and often after a long and entangled bargaining process in which they are unable to involve the local people they have organized. The victory, when and if it comes, doesn't look much different from any other welfare benefit, delivered by authority. At some point, it is hardly necessary to organize at all. One only needs legal standing or the political skills necessary to make city councils, regulatory commissions, and congressional committees listen. Then organizing as an activity passes over into pure advocacy, like the work of the Nader groups: cooperating with some officials, fighting with others, seeking general publicity (rather than speaking to a particular constituency), going to court.

At the local level—and always in union work—organizing and advocacy go together. There is even some room for direct action, as in the sit-ins at nuclear plant sites organized by groups like the Clamshell Alliance of New England. Indeed, the politics of the antinuclear movement, and of the environmentalists more generally, comes closest to recapitulating the style and ideology of the early New Left (and of the early civil rights movement). But even these groups, whose members defend the proposition that small is beautiful, regularly appeal to the state to assert and enlarge its powers. They want state officials, for example, to impose safety standards on corporate and entrepreneurial activities. And while the stringency of the standards is still the subject of political debate, few people would deny that officials should do that—for the sake of all of us who are hardly able to do it for ourselves. When activists act on behalf of the unorganized and the unorganizable, they are always, explicitly or implicitly, asking the authorities to do the same thing.

It is often said, however, that when the authorities agree and extend the reach of welfare or regulation, they turn the men and women for whom they act into helpless dependents— eternally posed, in contemporary right-wing literature, with their hands out. If there is any truth to that view it is only a partial truth (to which I'll return later on). The scale of corporate institutions and activities also imposes dependency, and of a more drastic sort, because the corporation is not even potentially subject, as the state is, to the control of its dependents. So people organize to resist its power, and the welfare state is created to mediate the struggle. The mediation is incomplete. It works well for the well-organized, especially for highly cohesive national constituencies. For the unorganized and the unorganizable, and commonly at local levels of government, it hardly works at all. But in principle, at least, it can be made to work. The state is accessible in ways the corporation is not. There are political entry points, even for individuals and groups who cannot bring economic power effectively to bear. Hence the role of organizers and advocates. They insist upon the full meaning and the serious enforcement of the welfare and regulatory codes. Their "historical task"—it's not quite what their ideology tells them—is to expand the scope of the welfare state, to make the state effective at the local level, in detail, and for everyone.

They work within the system—these alienated, embittered, "radicalized" militants of the sixties—because there is no other place to work. And the system is constraining. Victories are possible, but these are always, as I have said, small victories, and the work is demanding, difficult, frustrating. Moreover, it is not obvious that the victories add up. Local organizing does not lead in any clear way toward socialism—no more than union organizing did in the thirties (though the prospect seemed brighter then). Organizing and advocacy in the contemporary welfare state create clients—first of all for the organizers and the advocates, and then for the state—not self-determining men and women. This is the truth behind the neoconservative critique of welfare and regulation. But there is an easy response: clients are at least men and women for whom someone speaks. The greater danger today is to be unspoken-for.

IV

Among the old/new activists, there exists a certain view of their work that incorporates, or might incorporate, much of my own argument. They talk about "base building" or "softening the turf"—for they are literally in retreat, waiting for some new upsurge of leftist activity, a national movement to which they might deliver their clients and constituents. All that they can do right now, they say, is to break down existing barriers to radical argument and political identification among the great American majority: white workers, government and corporate clerks, young professionals, women. They need to win the small victories; they don't expect to win big ones. Indeed, some of them probably share the neoconservative view that the big victories are impossible. The politics of entitlement "overloads" the welfare system, and the system simply cannot deliver the benefits its rules and regulations promise, not if these are widely demanded. Hence at some point the demands will be blocked. Then it will be the task of the organizers to explain the nature of the blockage.

I am unsure that this view is right, or that it will be seriously tested in the near future; the American welfare state, for all its immediate problems, still seems to be a highly expandable system. Nevertheless, it is an intriguing view. Imagine a world where tenants successfully insist upon the minimal safety and amenity standards required by contemporary housing codes, where service workers of all sorts organize and defend their interests as effectively as teamsters and steelworkers, where women must be paid as well as men, where high safety standards are imposed on car manufacturers and power companies, where neighborhood alliances demand and receive equal services and equal protection—in short, where people get what they are "entitled" to. Is it possible? What would it cost? And who would pay?

To deny that it is possible is to suggest that the contemporary welfare state requires a large under-class, not only or most importantly of poor people, but more generally of unorganized, passive, and inarticulate people. Perhaps this is so. It may even be the case that no alternative (socialist) state could actually give people everything they are entitled to. So long as entitlement is understood in terms of the receipt of benefits, it has inherent limits—which are, at the outside, more or less similar whatever political arrangements are established. The "outside" is, to be sure, far away. Welfare politics remains the most important politics today, and no serious activism can avoid the patterns it imposes: advocacy and administration, individual entitlements and bureaucratic interventions.

But it is necessary, at the same time, to think about something very different. The good society will not be a republic of clients whose relation to the *res publica* is simply that they appropriate its resources under fair distributive principles. Nor would it be a satisfactory outcome of contemporary struggles if the welfare state were left, at the end, as the only effective political organization or as the universal mediator of social conflict. It will not be satisfactory because such a state would indeed be "overloaded," and because its beneficiaries are not likely to sustain for long the parts of democratic citizens. Somehow, the shift from organizing to advocacy has to be reversed, so that cohesive groups take shape whose members are not only consumers of benefits but active participants, capable of mutual assistance and even of mutual restraint. I am not sure how that reversal might be effected, but it is important that there be people at work not only at the center of the welfare state but also, so to speak, in the parishes. We should be sympathetic, then, to the pastoral retreat of the New Left. It represents both a surrender of millenarian politics and a stubborn recommitment to the Good Old Cause.

(1979)

PART III

Social Change

12

Modernization

I

Historical concepts always tell us as much about the people who use them as about the events they are supposed to describe. Very little is given in intellectual life; artists, writers, even social scientists, must choose the way they wish to see the world, and invariably they choose ideas which suit their needs or their convenience. American social scientists use the concept "modernization" first of all for reasons of convenience; it groups together all the complex transformations now going on in Africa and Asia, and it suggests useful comparisons with similar transformations presumed to have occurred at one time

I don't pretend in this essay to examine in detail the large number of books and articles in which the theory of modernization is used more or less systematically, nor the even larger number in which it is merely invoked. My purpose is only to suggest some of the intellectual and political tendencies fostered by the theory and already visible in the works of Daniel Lerner, W. W. Rostow, S. M. Lipset, Gabriel Almond, etc.—the writers who have most ably worked it out. No doubt, they disagree among themselves, and my criticisms by no means apply equally to each of them. But modernization, in theory as in reality, has acquired a momentum of its own. Its terms inevitably have come to have meanings for which its creators are not entirely responsible, even when these meanings are (as they often are) true to the logic of the theory.

or another in Europe. Modernization is thought to be a single, long-term historical process in which all mankind is destined to participate, but in which some men and women already have participated. Their experience provides the basis for a theory about the stages, crises, and turning points of the modernizing process, and about the character of its active agents and still vigorous opponents.

Exciting work has been done with this conceptual scheme, but the concept has serious difficulties and these have seldom been treated analytically. Nor has anyone suggested that there may be reasons beyond scientific utility for its extraordinary popularity. One such reason is probably that modernization theory forecloses a great deal of political debate. Even the casual use of words like "backward" or "developmental process" suggests forcefully that the direction of historical change is not at issue. Modernization theory thus opens the way for a social science as self-confident as was the historiography of nineteenth-century Whigs. American academicians seem as certain as were Whig historians that their contemporary world is the modern world, and the modern the very best world there has ever been.

"Modern" has become a term of self-congratulation; it is a way of indicating those aspects of the present in which present-day men and women take special pride and by which they distinguish their own cultural, political, or technological achievements from those of previous generations. It is not really a historical, let alone a scientific, word at all; it is a word, so to speak, with built-in naïveté, the characteristic naïveté of writers who make the contemporary and familiar into something superhistorical.

When social scientists accept this prideful evaluation of contemporary life, they cut themselves off from historical understanding. Their theory then fails to do what a theory of history and society ought to do: to suggest possible or likely connections between the past and future. "Modernization" is by definition a terminal process; it culminates in modernity.

It suggests connections only between the historical past and the unhistorical present, the conventionally admired world around us.

The theory of modernization appears in an earlier form as Weber's "rationalization"—a process culminating in a "rational-legal" society. Rational-legal and modern are terms very similar in meaning, but Weber avoided the word that his followers have chosen, perhaps because he realized the instability of the social system he was describing. Acutely aware of the discomforts of rational-legal civilization, he sensed that there would be future transformations. He warned of revolutions led by charismatic men, calling into question all the achievements of human rationality. But the American social scientists who write of modernization live wholeheartedly in the midst of the modern. Like Hegel in the Prussian state, though with a more catholic sense of human destiny, they view their own achievements as the culmination of historical development. Theirs is a theory of progress, but of *progress realized*. Unlike the eighteenth-century French doctrine, it does not point beyond the present. "Modern" men and women decline to speculate on their future, but absorb themselves instead with their heroic and painful past—confident that masses of people in other parts of the world are now re-enacting that past. Complacently they wait for their straining, backward fellows, as if they had no more history of their own to act out.

II

Marxist theory was constructed around the notion of two revolutions: one had already occurred when Marx wrote, though not yet in all countries; one had not yet occurred in any country. These two revolutions linked three stages of human history: feudal, capitalist, and socialist. The theory of modernization, by contrast, rests on the idea of a single revolution that has already occurred in some but not yet in all countries. This revolution links two historical stages: traditional and modern. The current doctrine, then, represents a major act of intellectual reduction, an act not without some empirical justification. For no country has had more than one revolution, or at any rate, no country has had two revolutions that can usefully be described as bourgeois and proletarian. Marxist writers have struggled to explain why it is that "proletarian" revolutions have occurred only in countries that have never had "bourgeois" revolutions. Theorists of modernization have no trouble in avoiding this difficulty, for in effect they argue that only one revolution is ever to be expected. That revolution marks the final crisis of the old order: a vanguard of modern men and women bursts through the trammels of tradition and establishes the new society. Because of the complex processes through which ideology and technology are imitated and diffused, however, this crisis occurs at different times in different countries and draws different groups into its vortex—businessmen, factory workers, peasants. The character of the emergent society will be determined by the point in economic development at which the crisis takes place and by the social classes which respond to its turmoil. That society can be *either* capitalist or socialist.

Modernization theorists (following Weber) tend to deny that there are significant differences between capitalism and socialism. The two are equally the products of the revolution against traditional order. Their sociological and psychological foundations are thought to be identical. They both require, for example, the same work ethic: thus recent writers have rediscovered Weber's Puritanism among the Bolsheviks. They both foster the conjugal family and shatter the ties of extended kinship systems. As against traditional passivity, they both generate mass activism and participation (though neither requires that that participation be meaningful or democratic). They both depend for efficiency upon an impersonal and bureaucratic rationality.

Marxist writers were not, of course, blind to these aspects of social life, but they did center their analysis upon the ownership of property, an issue that now seems secondary. For them, feudalism and capitalism, where property was privately owned, were more closely related to each other than either was to socialism, where private property was abolished. In other areas as well, however, socialism was thought to involve significant transformations precisely of modern, that is, bourgeois life. Socialist writers, for example, foresaw the transformation of work into a kind of play or, at any rate, into a creative, cooperative, and free activity; they thought that the Protestant work ethic with its overtones of compulsion was characteristic only of capitalist society. They looked forward to changes in the conjugal family, for the household with all the inhibitions it placed upon the freedom of women had clearly survived the breakup of the extended kinship system. And so on: they wrote also of changes in the character of political participation; they advocated the radical decentralization and simplification of bureaucratic structures. Yet none of this is integrally connected with socialism in modernization theory—nor, it need hardly be said, in the modern world as we know it. According to our theorists, all these proposals are merely the utopian effervescence of the crisis period, quickly enough forgotten when the time comes for socialists to take over governments and begin the hard but rewarding work of modernization. The theory thus predicts the failure of socialist aspiration insofar as it points beyond the modern world, even while admitting the possible success of socialism as a modernizing ideology.

Once again, there is historical justification for this view. The two characteristic forms of contemporary socialism can both be accommodated better by modernization theory than by Marxism. Social democracy seems indeed a modern social system, far closer to welfare state capitalism than to the Marxist vision. Similarly, Bolshevism is probably best described as a modernizing ideology, its adherents driven to act out that repressive and disciplinary role currently thought necessary to social progress. And the society that the Bolsheviks have produced certainly bears striking resemblances to the modern world that has emerged in the West—there also without the benefit of a second revolution. Socialists, in fact, have nowhere produced societies conforming to their own aspirations. In the name of progress and of the efficiency and bureaucratic rationality that progress seems to require, they have set aside one by one precisely those purposes whose achievement would have resulted in a radically new form of modernity and would also have required major changes in the theory of modernization.

If such changes are not required by anything that has yet happened, however, one can at least insist that they may yet be required. The aspirations expressed during the "crisis" period are indeed not fulfilled by modernity; but neither are they abolished. For this reason, the Marxist writer has one decisive advantage over the academic theorist of modernization. He is unalterably committed, living as he always does *before* the second revolution, to search in his own society for those "contradictions" in social, economic, and political life that are manifest in human discontent. He quickly becomes aware of the way in which historical problems, like those seemingly outmoded social classes in Marx's *Eighteenth Brumaire,* accumulate through time. He discovers the unevenness of historical change and senses the troubles of the men and women left behind. Most importantly, perhaps, he understands how historical unevenness fixes group interests and social structures into patterns of conflict that are extremely difficult to overcome. And he sets this overall view of society against the possibilities that the future holds and toward which the aspirations of actual people point. Now, none of these insights is in any way unavailable to the theorist of modernization so long as he studies pre-modern societies or societies that are in some sense in transition toward modernity. The theory of the single revolution, however, commits the social scientist who holds it to view the problems of "rising expectations" or historical unevenness as susceptible to definitive solution through the completion of the modernization process. As that process nears completion, he tends to talk of "areas" of backwardness and "pockets" of discontent and then to call for political or economic "mopping up." He is singularly blind to the anxieties and frustrations that seem endemic precisely in the "modern" life he unquestioningly accepts and that are the result, in part, of the very failure of socialist aspiration that he claims to understand. And he surely does not see the way in which the structure of modernity, like any other structure, may become an impediment to further progressive transformations.

III

A theory with one revolution is a theory with one revolution too few. However useful it is in explaining the past, it fails to alert the writers who use it to that which is transient in the present. But there is another difficulty in modernization theory: *it fails to provide any very interesting means of distinguishing among different presents*. All sorts of distinctions, some rather obvious and some very subtle, can be made among societies that are on their way to modernity, but it is hard to avoid the conclusion that modernity itself is singular in form. Modernization theory, then, lies behind the widely shared view that the United States and the Soviet Union are evolving toward one another and eventually will develop very similar social systems. This is a view that suits men and women of different political persuasions: it can provide the basis for arguments about Soviet liberalization or incipient American totalitarianism. The bias of the theory, however, would appear to favor a forecast of Soviet liberalization, both because the idea of the modern has been shaped in the image of Western experience, and because the Russians being less recently modern presumably have further to evolve.

If Bolshevism is seen as a modernizing ideology, then

Russian totalitarianism can be described as an institutionalized arrangement appropriate to a certain stage in the modernization process. Such a description has a certain plausibility, especially if one ignores or de-emphasizes the purges of Stalin's time, which seem without historical "appropriateness." Alternatively, totalitarianism, with the purges, can be conceived as a "disease" to which nations are peculiarly susceptible during the crisis period—a view that justifies efforts to prevent the disease and has therefore been favored by American writers. Now either of these views represents an advance over the Cold War conception of the Soviet system as permanent and monolithic evil. Both of them provide a historical framework within which totalitarianism can be better understood than it has been until now. At the same time, however, both of them do us (and the Russians) the acute disservice of suggesting that the future of Soviet society is somehow not a problem (only the past is a problem): the modernizing process will go on; the disease of Stalinism will be shaken off and the patient brought to a condition that, it is assumed, will be healthy simply because it will be modern.

Of course, totalitarianism is bound to be transformed as the historical process of which it is the product proceeds. The question is, transformed in what way? All processes arrive and are always arriving at the modern, but this is always a particular modern, determined by its history. Not even the spread of a single technology or the increasing interrelatedness of the various divisions of mankind seems likely to produce in the near future a single history and a singular modernity. We will continue to live in significantly different societies. And for the moment, at least, it is useful to argue that some of these societies will be totalitarian. That word need not retain its apocalyptic overtones; its use is only to suggest, in the language of modernization theory, that while the "crisis" may be comparatively brief, the particular ways in which it is met have lasting effects.

Eighteenth-century theories of progress were characterized

by a curious notion that the past might be abolished: each new advance in science or politics not only overcame the obstacles that previous generations had erected, but literally annulled the darkness in which they had lived. So men and women continually escaped the consequences of their own history. Something of this odd notion survives in modernization theory: at a certain point in the process of development, it is thought, individuals and nations loose themselves from their parochial pasts and, after one or another sort of social trauma, enter a universal present. There is some truth to this, perhaps, if it is expressed with sufficient caution. The immediate past is more parochial, the present more universal, chiefly because of the worldwide diffusion of ideas and artifacts, itself an aspect of modernization. But this very diffusion evokes from each country a particular response, leads to a special form of social unevenness and eventually to a distinct political creed and a cherished independence. The first result of diffusion is to increase the number of sovereign states, the number of arenas in which political decisions are made. And insofar as parties and politicians make different decisions, they or their descendants will find themselves living in different worlds. This remains true even after one admits that the range of choices is not always wide and that the word *choice* does not always indicate debate and deliberation.

The Russian people thus are likely to endure their modernizing party as an inescapable incubus for years to come, long after its presumed historical function has been fulfilled. Soviet modernity does not depend first upon the activity of the party and then upon its withering away. Instead, the party creates a particular form of modernity, a form to which it is itself integral and not incidental. Similarly, the Western experience of Puritan repression and bourgeois egotism is not simply one pathway into the modern world. Repression and egotism give rise to modern capitalism and are to an extent embodied in their own creation. They form the ever-present past and become our own incubi when we struggle to adjust to (or deny) the possibilities of abundance and automation. Neither the institutional structures of individualism nor the Communist party, of course, is a necessary part of modernity. On the other hand, one cannot discover the essence of modern life by excluding them from one's conceptual scheme. That excludes too much: modernity always appears in particular historical forms, in which Americans and Russians, Chinese and Indians struggle with the past, adjust to it, evade and endure it, but never escape it. When modernity is turned into a sociologist's ideal type, it suggests that impossible escape and then does not provide any clear way to analyze or understand the particular forms of the modern—where the encounter with the parochial past is as essential as is the universal present.

IV

These criticisms of modernization theory apparently have more relevance in the "advanced" than in the "backward" countries. The nearer one is to the fully realized modern, the more disturbing is the idea that radical social change is a thing of the past. The theory is most successful in explaining the experience of backwardness, that is, the new awareness of people living in traditional societies that they are in some sense "behind" and must quickly "catch up." Indeed, the major purpose of the theory is to facilitate the business of catching up. Large numbers of American social scientists have suddenly become defenders of social change—in the backward countries. They have developed a healthy sensitivity to the human misery upon which traditionalism rests. They approach the problems of development with a lively pragmatism which is often appealing. There is a "job-to-be-done" air about modernization theory and the theorists may well feel that the pressure of poverty and hunger is so great that there is no time for complicated analyses of modernity. Somber warnings about the inescapability of the past—and especially of that past which is being made today—must sound to them like so much superfluous moralizing. The activity of the modernizers requires a goal that can be assumed; modern life must appear as attractive as it is inevitable. Then the human price for rapid social change will be gladly paid; all costs must seem worthwhile if modernity itself is the purchase.

Most American social scientists, of course, do not believe that all costs are justified; they study modernization at least in part to learn how coercion can be minimized and totalitarian terror avoided. Thus they have tended to see modernization as a process whose *means* require further study and debate, but whose *ends* are given. In fact, however, the givenness of ends limits and inhibits the debate over means. Given the character of contemporary backwardness and the character of contemporary modernity, it is probably correct to argue that modernizing regimes will always be authoritarian and their methods always coercive, if not terroristic. All debate is then reduced to questions like, how authoritarian? how coercive? But this reduction is only necessary if the model of modernity imitated in the backward countries is not itself subject to change and if there is only one possible end to the developmental process. The alternatives in the backward countries depend, then, upon the alternatives in the advanced countries.

In theory, at least, a thoroughgoing critique of the modern world would offer men and women everywhere a wider range of choices. It is in this sense that the criticisms suggested above are relevant even to the politicians and economic planners in Asia and Africa who are "doing the job." So much of what they call modern is already outmoded, vestigial, inadequate to human needs! With automation so near a prospect, for example, how necessary to the job they are doing is the discipline of a "Protestant" work ethic? Or again, now that decentralization is technically more feasible than ever before, how necessary to that job is the creation of a bureaucratic hierarchy? So long as modernity is conceived in a singular and narrowly contemporary image, with "Protestant" workers and a centralized bureaucracy part of that image, there is only one historical pathway and it culminates in the American present—or in that immediate future that Khrushchev promises the Russians. Presumably, people must struggle along that pathway as quickly and with as little pain as possible, and since the end is preordained, theorists can only distinguish between forced marches and voluntary advance. If, by contrast, modernity is seen as having a multitude of forms, then there are a multitude of paths, and all sorts of choices to be made and opportunities seized.

But contemporary social scientists seem unwilling to confront the modern world in a critical way. They are content to view social problems as functions of backwardness and to offer modernization as a single cure. Thus they become social critics and even revolutionaries—abroad—out of sympathy for their fellowmen, in a good cause, and with admirable energy and skill. If only modernization offered some solution to the problems of the modern! And if only these writers were revolutionaries at home!

(1964)

13

A Theory
of Revolution

I

Most Marxist writing about revolution, by academics as well
as militants, has focused on the great question: how to get
started? What are the causes of revolution? There has been
less interest, surprisingly little, in outcomes. A certain agnosticism about outcomes seems to be a feature of leftist thinking, dating at least from the era of 1789. Thus, St. Just's
dictum, adopted by, and one would think more suitable to,
Napoleon: *On s'engage et puis, on voit.* Marx gave this agnosticism a historicist rationale, though only with reference to the
last or proletarian revolution. The world whose laws science
could discover and explain lay this side of that cataclysmic
event, and what lay on the other side was largely unknowable.
Or, rather, it could be understood only by negation—the
withering away of the state, the abolition of private property,
the achievement of classlessness—and not in any positive or
substantive way. "The dictatorship of the proletariat" re-

mained a phrase without content, and the political, administrative, and economic character of communist society was never seriously discussed.

Lenin and Trotsky laid the foundations for a theory of outcomes but did not develop it in any detail; nor did they acknowledge its political implications. To have done so would have undercut their own activity. And yet, they must have had some idea, before they committed themselves, of the sort of regime they would be creating. In any case, such an idea is implicit in their writings, and I shall try to expound it. I do not claim that what follows is an account of "what Lenin and Trotsky really meant." It is only one possible working out of an argument they began. I merely follow certain familiar clues, turning sometimes to historical examples that the Bolshevik leaders would never have chosen. The clues have to do, first, with the internal structure of revolution—with the sequence of events and the relations of forces within the process—and second, with two very different kinds of revolutionary endings.

The term *revolution* obviously does not cover every attack upon an established order or every seizure of power. Military coups are not revolutions; nor are most anticolonial struggles. In a world in which political turnovers are common, the term covers only a small number of cases: conscious attempts to establish a new moral and material world and to impose, or evoke, radically new patterns of day-to-day conduct. A holy commonwealth, a republic of virtue, a communist society—these are the goals revolutionaries seek. So I shall focus on the great revolutions—the English, French, Russian, and Chinese —in which modern radicalism reached its fullest substantive expression and the new world came most clearly into view. The argument about structures and endings is essentially similar in these four, whatever other differences exist among them. Now that we have seen Lenin's revolution, and Mao's, agnosticism is no longer a practical or a justifiable option. Nor, unhappily, have the most recent revolutions carried us into a world of freedom, beyond the grasp of social analysis.

II

Revolution, then, is a project, and it is important to say whose project it is. This is the question Lenin addressed in *What Is To Be Done?* When we study the forces that make or try to make a revolution, he suggests, we immediately discern two groups, with different sorts of political capacities and ambitions: a revolutionary class whose discontent provides the energy and whose members supply the manpower, and an intellectual vanguard that provides ideology and leadership.[1] The vanguard is formed only in part, perhaps in small part, by men and women drawn from the revolutionary class. The extent of recruitment depends largely on the social composition of the class, the availability of education to its members, and so on. Thus, a significant number of Puritan clerics came from the lesser gentry; an insignificant number of Chinese Communist intellectuals come from the poorer peasants. By and large, while classes differ fundamentally from one revolution to another, vanguards are sociologically similar. They are recruited from middling and professional groups. The parents of the recruits are gentlemen farmers, merchants, clerics, lawyers, petty officials. Recruitment begins at school, not in the streets, or in shops and factories, or in peasant villages.

Lenin argues, though explictly only for the proletariat and the Marxist intelligentsia, that each of these groups has its own consciousness. Class consciousness develops as the spontaneous assertion of the shared interests of the members, as these interests are perceived by men and women still living in the old order and still thinking only about its possibilities. They have little choice; they are ambitious but hemmed in, or hard pressed simply to survive. They have to make out, or they have to earn a living today and then tomorrow. Their

[1] *What Is To Be Done?* (Peking, 1975), esp. Part I; Lenin draws freely here upon arguments first developed by Kautsky.

shared awareness of their predicament moves them to associate for protection and short-term advance. Hence the parlia-
mentarianism of the English gentry and the trade-unionism of
the modern working class. Though the life patterns of the
revolutionary class may point toward a new social order, its
conscious activity is shaped within the old order and aims
at accommodations thought to be possible. Class consciousness rarely inspires an innovative politics. The idea of radical
transformation is carried into the revolutionary class by the
men and women of the vanguard.[2]

Vanguard consciousness is the work of intellectuals somehow cut loose from the constraints of the old order—or of
intellectuals who cut themselves loose. These are people,
usually young people, who respond to the decadence of their
world by withdrawing from it. They give up conventional
modes of existence, conventional families and jobs; they
choose marginality; they endure persecution; they go into
exile. They are receptive to radical and, as their opponents
rightly say, to foreign ideas: Calvinism in sixteenth- and
seventeenth-century England, English liberalism and Genevan
republicanism in eighteenth-century France, Marxism in Russia
and China. Revolutionary thought nowhere develops indigen-
ously. Nor does the will to revolution—at least, it does not
arise in the center but at the furthermost edges of the pre-
revolutionary world. If the new class grows in the womb of
the old society, its delivery ("Force is the midwife . . .")
comes from outside.

Class and vanguard consciousness are very different, and in
characteristic ways. An analogy with the Israelite Exodus from
Egypt, used frequently by Puritan radicals and occasionally
still by the Jacobins, illustrates the difference. A double consciousness guides the Exodus. The people (the oppressed and
revolutionary class) are moved by the vision of "a land of
milk and honey"; Moses and the Levites (the vanguard) are

---

[2] I have tried to illustrate this thesis with regard to the English case in
The Revolution of the Saints (Cambridge, Mass., 1965), chap. 4.

moved by the vision of "a nation of priests and a holy people."
Both these groups and both these visions are necessary for
success. Without the people there would have been no new
nation; without Moses and the Levites the land would never
have been conquered. As the biblical account makes clear, the
people alone would probably not have left or would quickly
have returned to the fleshpots of Egypt. It would be wrong to
think about this as a simple conflict between popular interests
and intellectual idealism. For it is possible to be very idealistic
about milk and honey, and groups like the Levites quickly acquire a vested interest in holiness. Each side has interests and
ideals which overlap in complex ways and make cooperation
between priests and people, vanguard and class, possible. The
two forms of consciousness reflect two different experiences—
that of the slaves in Egypt, that of Moses in exile in the
desert—which are nevertheless the experiences of people tied
to one another and capable, at some level, of understanding
one another.

The two different experiences produce two different sorts
of political association. Class politics is catholic and inclusive.
Gentlemen, merchants, workers, and peasants in the old order
share a common life—share experiences, willy-nilly, without
reference to the opinions or feelings of individuals. It is a
matter of collective location, not of private volition. Hence,
class organizations are open, and their internal life usually
takes shape, at first, as a democracy of the members, loosely
governed.

Vanguard groups, by contrast, are closed and exclusive.
Joining the vanguard is a matter of choice, but it requires also
the acceptance of new members by the old. And the old have,
through choices of their own, established certain criteria.
They need to make sure of the commitments of their would-be
brethren or comrades in order to guarantee the special character of their group. Here, collective location or class origin
hardly matters. What does matter is opinion, ideology, zeal,
readiness to accept a common discipline. Crane Brinton has

said that Jacobin ideology amounts to a call for a nation of smallholders and shopkeepers, "a greengrocers' paradise."[3] That is true, up to a point, and one would expect all greengrocers to be welcome. The Jacobin intellectuals, however, had in mind a republic of virtuous greengrocers—tested and certified at meetings of the Jacobin clubs. The two notions overlap but are not the same.

The inner history of the revolution is in large part the working-out of the tension between these different notions and between the two groups of men and women who carry them. "The shift in different stages of the revolution," Trotsky has written, "like the transition from revolution to counter-revolution, is directly determined by changing political relations between . . . the vanguard and the class."[4] These relations in turn are shaped, I shall argue, by the different social compositions and the relative political strengths of the two groups. The thrust toward revolutionary dictatorship, the pursuit of holiness, virtue, or communist discipline, the use of terror, the possibility of a Thermidorean reaction, success or failure in the establishment of responsible government—all these depend upon the interaction of vanguards and classes and then on the historical factors that determine the interaction.

The analysis cannot begin, then, with either vanguards or classes considered alone, for what is crucial is the relation between a particular vanguard and a particular class at a particular moment in time. The balance of forces, the relative strength and competence of the two groups, shapes the revolutionary process. Ideally, the balance should be described in careful detail, but I obviously cannot do that here. I can only offer a quick historical survey of class/vanguard relationships in the great revolutions.

A clerical vanguard, like that of the Puritan ministers, holds a strong position over and against any lay group. It stakes a claim to special knowledge, though no longer to magical

[3] A Decade of Revolution: 1789–1799 (New York, 1934), p. 136.
[4] "Hue and Cry Over Kronstadt," New International 4 (1938): 103.

powers, and it possesses a considerable capacity for collective discipline. This capacity was evident when young and radical clerics established the first underground organizations in modern European history. Protestant ministers, however, are vulnerable to the appearance of lay saints, who may either join and reshape the vanguard or organize to resist its initiatives. The appearance of born-again Christians among gentlemen and merchants quickly undercuts the more extreme forms of clerical pretension. Lay vanguards, led most often by lawyers and journalists, hold a weaker position relative to gentry-merchant groups, for their knowledge is not so special and is shared almost from the beginning by the men and women with whom they interact. It is among these middling social groups that intellectuals are most likely to fulfill the task that Lenin first assigned to them—which he and the Bolsheviks were never able to fulfill: "The task of the intelligentsia is to make special leaders from among the intelligentsia unnecessary."[5] Hence, the weakness of radical intellectuals as a distinct and disciplined group during the French Revolution and the virtual nonexistence of vanguard politics in 1830 and 1848.

Organization requires more than competence, however; it also requires practice. The hundred years of Protestant experimentation with conferences and congregations—roughly the period of gentry self-assertion in Parliament—goes a long way toward explaining the precise form of the interactions of the 1640s and the 1650s. In the years immediately following 1789, by contrast, the radicals were compelled to innovate on the spot. Eighteenth-century French society had only the most rudimentary sorts of lay political or intellectual organization (the salons, literary and scientific societies, Masonic lodges). The Jacobin clubs, split and purged several times, represented a first approximation to the party cells that facilitated later vanguard activity. But they lacked trained and disciplined cadres and members sufficiently differentiated by experience

[5] *What the "Friends of the People" Are* (Moscow, 1951), p. 286.

or conviction from their immediate social surroundings. The short life of the Jacobin republic, and its failure to leave any significant institutional residues, has to be connected with the short history of Jacobinism before the republic was founded.

In more recent revolutionary upsurges the independence of the vanguard has been enhanced by its contact with poorly educated and unorganized social classes. Vanguards have a much stronger position relative to new industrial workers and traditional peasants than to gentlemen and merchants, and a stronger position relative to peasants than to workers. Selig Perlman has argued, also using a Leninist sociology, that the power of a radical intelligentsia within or over the working class declines in close connection with the rise of unionism.[6] The more organized the class, the less powerful the vanguard. If that is so, then it follows that the proletariat of a developed industrial society will resist vanguard initiative more strongly than other social groups, and for what may properly be called Marxist reasons: everyday life tends to produce among workers very high levels of solidarity and political sophistication and relatively tight defensive organizations. Perhaps that is why there has never yet been a revolution in which a mature working class provided the mass base.

The conditions under which social classes yield to vanguard direction resemble those that make for other sorts of elite dominance. First, class balance: that moment in history when an older ruling class can no longer maintain its political position, while the coming class cannot yet assert its own authority. Engels refers to this balance of forces in explaining early modern absolutism.[7] He might have added that it helps explain the role of radical intellectuals in the struggle against absolutism. Once that struggle has begun, however, the balance is likely to shift rapidly toward the rising class, and if that class is sufficiently cohesive and well organized, the van-

---

[6] *A Theory of the Labor Movement* (New York, 1928).
[7] *The Origin of the Family, Private Property and the State* (Moscow, 1952), p. 281.

guard upsurge will be brief. It can be prolonged only if no social class can assert its own right to rule. Second, then, class underdevelopment: the essential prerequisite of sustained vanguard dictatorship. A wide variety of factors come into play here. Class size, resources, education, organizational structures, and traditions of struggle determine the specific revolutionary capacity of burghers and proletarians. Mass illiteracy, geographic dispersion, a purely local solidarity determine the general incapacity of a traditional peasantry.

Modern radicalism has tended to reach out for a peasant base—to bring into political life a social class far less capable of organization and independent activity, more in need of and more at the mercy of vanguard leadership than any other. Not that the vanguard is ever entirely free of class control: I only want to suggest a comparative judgment. A Puritan minister was locked into a tight connection with, in part a dependence on, the English gentleman. Every move he made had to be negotiated. He had very limited powers of experimentation. Even when he succeeded in getting Puritan morality enacted into law, he could not get it enforced. The seventeenth-century gentry provide a classic example of a group resistant to vanguard initiative. Already in control of the Commons, politically sophisticated, well educated, economically powerful, it and its merchant allies were the agents of the first and perhaps the most successful Thermidorean reaction. Even this class, however, needed the clerical vanguard, at least for a time. For the ministers provided the decisive innovations in revolutionary politics and the zeal without which the monarchy could never have been overthrown.

It is easy to imagine how much more such men are "needed" when they interact with a disconnected class radically deprived of resources, education, and leadership. Peasants can mount Thermidorean pressures, such as those that forced the introduction of the New Economic Policy in Russia. But it is critically important that NEP was a Bolshevik policy aimed at appeasing the peasants, not a peasant policy aimed at overthrowing Bolshevism: "The concessions to the Thermidorean mood and tendency of the petty-bourgeoisie," Trotsky wrote in 1921, ". . . were made by the Communist party without effecting a break in the system and without quitting the helm."[8] There was no Russian (as there is no Chinese) equivalent of the gentry or of the French bourgeoisie, no indigenous class capable of generalizing its own way of life, asserting its ideological and organizational supremacy, and replacing the vanguard regime.

Thermidor decisively tests the class/vanguard relationship, and I should say a word about its general character. It is not to be identified too literally with the political intrigue of the summer of 1794. What made that intrigue possible was the disaffection of revolutionary forces from the Jacobin dictatorship and the widespread sense of alternatives short of a restoration of the *ancien régime*. Thermidor is not a counter-revolution, though it may open up possibilities for counter-revolutionary politics; it is rather the self-assertion of the revolutionary class against the politics of the vanguard.[9] If, in the Russian case, the proletariat is the revolutionary class, then Kronstadt and the Workers' Opposition represent failed Thermidorean tendencies. If we prefer the peasants, then NEP is as close as Thermidor ever came. The politics of the vanguard shapes the period of revolutionary history called—the name is a triumph of antivanguard feeling—the Terror. This term too should not be identified in any simple sense with the proscriptions and judicial murders of the Jacobin regime.[10] The Terror is the dictatorial imposition of vanguard ideology. So Thermidor marks the end of dictatorship, and its success or failure is determined by the "changing political relations"

---

[8] Quoted in Isaac Deutscher, *The Prophet Outcast* (London, 1963), p. 317.

[9] See Georges Lefebvre's summary statement in *The Thermidoreans* (New York, 1966), chap. 11.

[10] But Deutscher was surely right to argue, against Trotsky, that the Stalinist purges were a feature of the Russian Terror and not of the Russian Thermidor: *Prophet Outcast*, p. 316.

of the vanguard and the class. If Thermidor fails, the Terror becomes permanent.

III

Vanguard ideology, and therefore the political character of the Terror, has a similar form in each of the great revolutions. Its specific content reflects shifting intellectual traditions— Reformation theology, neoclassical republicanism, Leninist political theory. But the basic structure and the general themes of vanguard argument persist, even when radical intellectuals address different social classes. For the placement of the intellectuals with reference to the different classes, and to the old order as a whole, remains fundamentally the same. The presentation of the argument changes, of course, and it is probably worth noting a general decline in the intellectual quality, and a restriction in the referential range, of revolutionary literature from the seventeenth to the twentieth centuries. (I refer here only to the propaganda of the vanguard, not to its more reflective and theoretical work.) The rigorously argued and heavily annotated sermons of the Puritan ministers have no analogues in the popular writings of contemporary Chinese Communists. The debased aphoristic style —as it appears in English—of the *Little Red Book* is not imaginable in either the English or the French revolutions. Calvinism found popular expression at a fairly high level because the social classes for which it was popularized already possessed a substantial literary culture. Marxism has found popular expression at a low level because its vanguards have written for classes without a literary culture of their own.

But all such differences are less important than the deeper similarities in vanguard ideology. It is the similarities that make revolution the sort of event it has consistently been. The first is simple enough. Calvinism and Marxism, and republicanism too, though to a far lesser degree, impose upon their adherents a genuine intellectual regimen. Each of these creeds has behind it a tradition of learning; it requires study; and when studied it imposes order upon a wide range of historical, cultural, and political phenomena. Life in the vanguard is an educational experience. Its members come to possess a doctrine that they apply and manipulate with great skill, and this possession is crucial to their bid for power. Members of the revolutionary class remain doctrineless until they go to school with the vanguard. They share socially widespread interests and aspirations and hold common opinions; they do not need a doctrine. But the vanguard intellectuals, socially disinterested and often disdainful of common opinion, are likely to be, perhaps need to be, doctrinaire. Their zeal is first of all intellectual in character.

That zeal takes three different though related forms:

1. It is puritanical zeal. Vanguard ideologies place an enormous emphasis on self-control, collective discipline, and mutual surveillance (what the Puritans call "holy watching"). They aim to produce high standards of methodical work and to curb or limit all forms of the dissipation of energy through play. The vanguard's ideas about the range of revolutionary activity are likely to be fixed by their ideas about the number of people capable of sustained discipline of this kind. The range widens, at least in theory, as workers and peasants enter the revolutionary coalition. Thus, Lenin's *State and Revolution* suggests that everyone can internalize self-control and make it habitual, and that everyone can watch everyone else, so that policemen, like vanguard intellectuals, will one day become superfluous. But Lenin is not willing to wait for that wonderful day: "No, we want the socialist revolution with human nature as it is now, with human nature that cannot do without subordination [and] control. . . ."[11] The vanguard will have to do the controlling—there is no one else. In prepar-

[11] *State and Revolution* (New York, 1932), pp. 42–43.

ing its members for this task, Lenin often sounds like the prophet of a Weberian Protestantism, preaching vehemently against "this slovenliness, this carelessness, untidiness, unpunctuality, nervous haste, the inclination to substitute discussion for action, talk for work, the inclination to undertake everything under the sun without finishing anything. . . ."[12] For Lenin, the true sign of a "proletarian" was not his class background or his specific relation to productive forces, but his self-control.

Exactly how well self-control works, we do not know. It has not made policemen superfluous, but clearly there have been periods in the history of every vanguard when holiness, virtue, or Communist discipline has been maintained at a high level. And clearly too, the class with which the vanguard interacts always resists this difficult morality. Not, by any means, unanimously: the vanguard does make converts, especially during moments of crisis and heroism; modified versions of virtue do fit the experienced needs of particular classes or sections of classes; and the enforcement of morality can be turned into an expression of existing class conflicts (especially when the old ruling class was leisured, aristocratic, and "decadent"). So vanguard intellectuals find gentry, merchant, worker, and peasant collaborators. Nevertheless, the history of every revolution is in part the history of popular resistance to virtue. Once again the Exodus story is illustrative. According to a folk legend, on the day that the Israelites left Mt. Sinai, they marched at double speed in order to get as far away as possible. They wanted no more laws.

2. Vanguard ideology expresses a zeal for political activism and participation, for self-government, often understood as a consequence of or a parallel to self-control. It is almost as if a dialogue were going on between the autocrats of the old order and the vanguard of the new. The autocrat, like a good Hobbist, says, "Absolute power is necessary to social order;

[12] *How to Organize Competition* (Moscow, 1951), p. 62.

therefore I will repress you." The vanguard intellectuals reply, "Holiness or virtue or Communist discipline is necessary to social order; therefore we will repress ourselves." Self-government, understood as collective self-repression, must be the work of men and women who have already learned the discipline of the self, or who are locked into groups that enforce such discipline. Hence, the role assigned in vanguard political practice to the congregation, the club, and the party. By contrast, members of the class generally seek a more immediate form of self-government in parliaments, assemblies, and soviets. Hannah Arendt is right when she claims that freedom is the very essence of revolution.[13] But two different kinds of freedom are at stake in the revolutionary process. For the class, freedom is a natural or a human right already possessed; all that is necessary is to create conditions under which it can be exercised—to open up arenas for democratic politics. For the vanguard, freedom has to be earned; men and women become free over a long period of internal (religious or psychological) and external struggle.

Vanguard intellectuals, therefore, willingly repeat the arguments of older elites: that the people are not ready for self-government, free elections, a free press. Ever-present Thermidorean pressures, called counterrevolutionary by the vanguard, prove the unreadiness. The classic response to these pressures is the purge, which clears the political arena of men and women who, it is alleged, would vote if they could to return to Egypt. But the purge is ostensibly a temporary measure—permanent only for those who are killed in its course. Ultimately, and in principle, the band of brethren, citizens, comrades ought to include everyone.

3. The vanguard has deep egalitarian tendencies. Its activity calls into question all conventional social distinctions. Its members make war upon traditional hierarchies of birth and blood and denounce all claims to rule based on wealth rather

---

[13] *On Revolution* (New York, 1963), p. 21 ff.

than virtue. A hatred of personal dependency, a strong sense of the value of individual effort—these constitute the central features of every vanguard ideology. They are connected with the voluntarist character of the revolutionary struggle, and they are given symbolic expression in the new titles assigned to all participants, whatever their social background. Revolutionary classes, though they share for a time the excitement of the struggle, aspire to something simpler and less demanding. Consider, for example, two titles from the era of bourgeois radicalism: mister and citizen. The first reflects class consciousness and is nothing more than a demand for equal respect. The second reflects vanguard consciousness and, far more heavily loaded, implies a shared concern and a shared activity. And it is not clear, in the second case, how much respect there can be if shared concern and activity do not in fact exist.

When the vanguard reaches out to radically oppressed social classes, its moral egalitarianism generates also a commitment to material equality. Puritans and Jacobins never seriously challenged the property system, though the political prerogatives of private ownership would surely have been eroded in a holy commonwealth or a republic of virtue. Communist vanguards obviously go much further, though they in turn stop short of acknowledging the prerogatives of collective ownership. They do not yield power to the workers and peasants whom they recognize as equal owners of the means of production. Indeed, worker and peasant attempts to give their new equality a political form are likely to be treated as counterrevolutionary, especially after Communist vanguards take over the task of economic development.

In the earlier revolutions development was not a political issue, for the expansion of the economy resulted from the freely chosen activities of individual members of the revolutionary class. They required the support of the political authorities, but not their positive direction, the coercion of others, not of themselves. The case is very different in both

Russia and China, where the liberation of class energy and the assertion of class interest could not readily have produced anything more than an equality of the impoverished (in vanguard ideology—Maoism may be an example—an equality of the virtuous poor).[14] Hence, the vanguard is driven to take on the role played, under radically different circumstances, by the Western bourgeoisie. It generates hierarchical structures of roughly the sort that exist in advanced bourgeois society, though with different ideological justifications and disguises. Liberty and equality disappear from its creed. What is left is a commitment to the forms of self-control and labor discipline necessary for industrialization.

Thus, Lenin in 1918: "The task which the Soviet government must set the people in all its scope is—learn to work."[15] Here, government and people have replaced vanguard and class, but the relation of the first two is fixed by the relation of the second two. The government determines the tasks of the people, not the other way around. Once again, this pattern is possible only in the absence of an economically independent and politically advanced social class. The commitment to industrialization is also rooted in the vanguard's desire to maintain its new political position and to strengthen and develop the country it has come to rule. Vanguard intellectuals now seek to serve the long-term interests of their subjects. At the same time, they must ignore or repress the immediate demands of those same subjects. They are at war with "backwardness."

In the course of this war the members of the vanguard become more and more like the members of other ruling groups (modernizing elites?), increasingly accustomed to the prerogatives of government, increasingly isolated from their own people. Hence, the process that follows upon the seizure of

[14] See Benjamin I. Schwartz, "The Reign of Virtue: Thoughts on China's Cultural Revolution," *Dissent* 16 (1969): 239–251.
[15] *The Immediate Tasks of the Soviet Government*, in *Selected Works* (New York, 1935–1937), vol. 7, p. 331.

power might be called—the term is obviously not Lenin's or Trotsky's—the routinization of vanguard consciousness. But routinization in this case can be a harrowing business, for the militants of the vanguard carry over into their new bureaucratic roles a deep conviction of their own superior understanding (correct ideological position), a contempt for their enemies, and a disciplined readiness for combat. And they are, in the short run at least, steeled against the temptations of sentimentality and corruption.

IV

The argument thus far has assumed a single, readily identifiable vanguard and a single revolutionary class. Actual experience has been far more complex. The group of vanguard intellectuals shades off on the one side into the more responsible leadership, the directly controlled agents of the class (e.g., trade-union officials), and on the other into the exotic world of "new notionists," isolated sects and eccentric geniuses, without any social base at all.[16] The revolutionary class is itself a plurality of groups, perhaps a plurality of classes, including rising and falling, modern and traditionalist elements, not gentry, merchants, workers only, but artisans and peasants too. These different elements form an unstable coalition and come into conflict with the vanguard on different schedules; they also come into conflict with one another.

The revolutionary world, then, is more pluralist than I have suggested. And yet, it also yields regularly to an act of personal unification—to a dictatorship different from that of the vanguard, the dictatorship of a leader who seizes upon the disruption and disorder of the moment. The leader imposes on the revolution something of his own character, but he also

[16] The phrase is quoted in Christopher Hill, *Milton and the English Revolution* (New York, 1978), p. 108.

reflects the dominant tendencies of his society. Sometimes, he accommodates himself to the rising class and presides over the Thermidorean reaction, as Cromwell did in his last years. Sometimes, he intensifies and personalizes the Terror, as Stalin did before and after World War II. As with vanguards, so with dictators: their power is greatest where the mass base of the revolution is least organized and cohesive. The cult of personality grows where class political culture is underdeveloped. But why, in these circumstances, vanguard power cannot be sustained on a collective basis remains unclear. It is as if the radical intellectuals, for all their zeal and discipline, share in the political underdevelopment of their society—and if they do share that underdevelopment, they are likely, as in the Soviet Union, to suffer its consequences.

Still, personal rule is probably a temporary condition, and when we try to appraise the long-term outcomes of revolutionary activity, the class/vanguard scheme resumes its central importance. We can now distinguish two different sorts of outcomes. First, the vanguard wins and holds power, making its dictatorship permanent, dominating and controlling weak social classes. It attempts for a while to act out its radical ideology but undergoes a gradual routinization. Leaving aside the precise history and character of the routinization, it is fair to say that this was the foreseeable outcome of the Bolshevik revolution. The dictatorial rule of the vanguard was determined by the radical inability of any social class to sustain a Thermidorean politics. Thermidor, then, represents the second possibility: the revolutionary class resists and replaces the vanguard and slowly, through the routines of its everyday life, creates a new society in its own image. It reabsorbs the vanguard intellectuals into the social roles occupied by their parents, that is, into professional and official roles without any special political significance.[17]

The second of these seems to me the preferred outcome.

[17] See Crane Brinton, *The Jacobins* (New York, 1950), p. 230 ff. for an account of the later activities of Jacobin militants.

For popular resistance to vanguard ideology, even when it is unsuccessful, has been sufficiently emphatic and so often reiterated as to demand serious attention. One of the central features of the revolutionary process, it determines what we can think of as a revolutionary law: *no vanguard victory is possible without radical coercion.* Given that law, it is best to insist, if one can, and as early as one can, upon the superfluity of the vanguard. The best revolutions are made by social groups capable of articulating their own collective consciousness and defending themselves against the initiatives of radical intellectuals. Thermidor is the work of such groups—an optimal outcome since it generates a limited and socially responsible government, more or less democratic depending upon the size and confidence of the newly dominant class. Thermidor represents the fulfillment of Marx's vision of revolutionary politics—the moment when power is "wrested from an authority usurping pre-eminence over society itself, and restored to the responsible agents of society."[18]

Short of Thermidor, only two other possibilities might raise similar hopes. We might imagine an absconding vanguard, which withdraws from political power even in the absence of overwhelming class resistance. Like Machiavelli's ideal prince, it founds the republic, the new moral world, through its own heroic efforts, but then it "confides the republic to the charge of the many, for thus it will be sustained by the many."[19] Or, we might imagine a vanguardless revolution, carried out by a social class free from any lingering attachment to the old order, with a fully developed sense of its own future, capable of producing leaders of its own, loyal to itself—Marx's (but not Lenin's) industrial proletariat. But we have, as yet, no experience of such a vanguard or of such a class.

[18] *The Civil War in France*, in Marx-Engels, *Selected Works* (Moscow, 1951), vol. 1, p. 472.
[19] *The Discourses*, I, ix.

V

Let us, nevertheless, imagine such a class, for it is at least imaginable; its members would require only a strong and sophisticated sense of shared interests and a collective idealism. The absconding vanguard, on the other hand, belongs to the realm of political mythology, for it would require an almost saintly self-effacement, radically unlikely from men and women capable of seizing power in the first place. What might a vanguardless revolution be like? I will describe it in the future tense, foregoing the conditional, though I don't mean by that to make predictions about the future. It has to be stressed, first, that vanguardless does not mean leaderless. It only means that the leaders of the revolution will not form a closed ideological group, responsible to one another and to no one else. They will be responsible to the men and women they lead; they will be agents and representatives; they will co-exist within the movement or party with oppositional elements; their authority will be temporary and revocable. Leaders of this sort might still be brilliant, outstanding, capable of bold initiatives, but they will share the consciousness of their followers—express it more coherently, perhaps, or bring it to a point in a more dramatic way, but share it still.

Of course, there will also be radical intellectuals with their own consciousness and their own visions of possible futures; I would not wish them away. And the intellectuals will still form groups of their own: clubs, sects, parties, and editorial boards. But these will be ginger groups, attached to the larger movement in one way or another—generating new ideas, stirring things up—but unable to control it. Barred from conspiracy (not so much by the police as by the strength and solidarity of the new class), they will be forced to argue, persuade, and exhort; they will be limited, that is, to those activities that even they might one day come to regard as morally and politically appropriate to the intellectual vocation. Small numbers of them might break away for political adventures on the side, so to speak, acting out the old arrogance. But the more cohesive the class, the more brief these will be. As writers and teachers, the intellectuals will have some influence on class consciousness—and that will probably be all to the good—but they won't be able to replace that consciousness with their own—and that will certainly be all to the good.

But if class consciousness, as I have already argued, takes shape within the old order and aims in its politics at accommodations thought to be possible, how can it ever form a revolutionary creed? Assuming that particular sorts of success are in fact possible, I don't think it can be doubted that a vanguardless revolution will be a gradual movement, a "long march." It will take the form of a succession of accommodations in each of which the new class will find larger scope for its political activity and an increasing cultural influence. Without ever intending a total transformation, its members will slowly make their own way of life, their daily routines, into the common way of life. And one day, they will find themselves inhabiting, let's say, a workers' republic. They will have taken over or decisively reshaped the forms of everyday work and the means of production and then generalized that takeover in other areas of social and political activity. The process will develop in stages, but it is unlikely to be staged. It will include moments of tumult and upheaval, but it almost certainly won't culminate in anything like a one-stroke seizure of power.

All this suggests a pattern of revolutionary transformation appropriate to the highly industrialized countries; only there is the work force sufficiently skilled and sophisticated, organized and disciplined for the kind of politics that would be necessary. And even there, it remains an open question whether the workers—or anyone else—can in fact sustain the long march. Won't they get tired? Won't they get bored? Won't they be tempted, some of them, to join some vanguard adventure? Or to stay at home and let a new elite—recruited from their own ranks, perhaps, but no better for that—take over the movement? Certainly, Marxist writers have had intimations of such outcomes. One of Trotsky's close associates, Jean Vannier, wrote of the working class in the aftermath of World War II that[20]

It has shown itself capable of outbursts of heroism, during which it sacrifices itself without a thought and develops a power so strong as to shake society to its very foundations. . . . But by and by, whatever the consequences of its action, whether victory or defeat, it is finally caught up in the sluggish quotidian flow of things. . . . Its courage and self-sacrifice are not enough to give it what, precisely, is needed in order to act out the role assigned to it by Marx: political capacity.

The quotidian flow of things, the demands of the everyday: here is the deepest source of on-going political subordination. Vannier's lines suggest the inevitability of vanguards and then of bureaucratic elites, and the permanent possibility of terrorism. But we should never be too quick to agree to such suggestions. For there are ways of institutionalizing Thermidorean politics, not only in the state, but also in the movement, forms of political life that impose restraint and responsibility: periodic elections, oppositional activity, freedom of speech and assembly. Most Marxist writers have radically underestimated the importance of such things, treating them as if they were merely mechanical arrangements, useful or not at any given time. In fact, they are vital all the time; they are the beginning and the end. If the revolutionary movement is to create a democratic society, its advance must be an expansion from a center already democratic. Open membership, internal freedom—the forms of the future must be routinized

[20] I owe this quotation to Irving Howe: see his *The Critical Point* (New York, 1973), p. 18.

in the present, so that as the routines spread, the forms take hold, reinforcing political with economic self-government.

The test of a "rising" class, then, is its ability to maintain democratic procedures in its own organizations. It probably has to be said that the Western working class has not passed that test. Nor, however, has it failed and succumbed (like the masses of Russians and Chinese) to vanguard leadership. By and large, it has been governed by its own bureaucrats, and that is a government less and less stable as educational levels rise within the working class and as more educated (white collar) workers are organized. So the future is still open, and the vanguardless revolution, as I have described it, is still an imaginable process.

(1979)

14

Intellectuals to Power?

When Shelley wrote that poets are "the unacknowledged legislators of the world," he was not speaking in metaphor. He meant that they really do discover, shape, and announce the moral law. But Shelley was not making a political claim. He did not mean that the power of poetry should be publicly recognized or that poets should occupy the offices of state. In its most general form, however, that latter claim is common enough: that the state should be ruled by its most visionary, or at least by its most intelligent, citizens, and that they should rule, not by accident or luck, but because of their vision and intelligence.

Mostly, until now, other criteria have prevailed, and it has been the well-born, the powerful, and the wealthy who rule the rest of us. Poets and intellectuals have been politically successful in their own right only when they appeared as religious leaders or rode the crest of revolutionary movements. In secular dress and in ordinary times, without magical powers or doctrinal authority, they have been pushed to the

A review of George Konrád and Ivan Szelényi, *The Intellectuals on the Road to Class Power* (New York, 1979) and Alvin W. Gouldner, *The Future of Intellectuals and the Rise of the New Class* (New York, 1979).

sidelines. Sometimes, of course, the well-born and the wealthy
are themselves poets and intellectuals, on the side, but they
don't rule because of their writing or thinking; they make
other claims on our attention. Or, they simply surround themselves with poets and intellectuals, having learned to enjoy
the more cultivated forms of flattery. The poet-as-courtier is
a common figure in the *ancien regime*; the intellectual-as-
advisor is even more common in modern regimes. Servants of
power, sometimes lackeys: it is a position that breeds contempt on the one hand, resentment on the other.

But today, other feelings are apparent: admiration on the
one hand, pretension on the other. We live in societies that
produce extraordinarily large numbers of educated men and
women and that increasingly need their authority and decision-making skills. We even produce more poets (I think)
than any previous society ever did. It's not the poets, however,
who are leading "the march to class power," though they are
among the marchers, and already taste—so these two books
suggest—the state grants that will be theirs when the goal is
reached. The leaders are the more specialized seers of a secular age: masters of ideology and technical experts. Ideologists
and experts don't claim to rule because of their birth or blood
or land or wealth, but solely because of their insight. They
penetrate the complexity of modern economies and technolo-
gies; they have a grip on the historical process; they make predictions about the future. Theirs is a new legitimacy, and it is
not easy to challenge.

Insofar as the modern state is committed to planning, welfare, and redistribution, it plainly requires a vast civil service
of educated people; intellectuals are its natural rulers. And
so, a certain sort of common sense suggests, intellectuals have
set about to create such a state, whose offices only they can
fill. By and large, they have succeeded, or they are in the process of succeeding. Here then are our new masters: bureaucrats, technocrats, and scientists; and their professional allies,
doctors, lawyers, teachers, and social workers. There are

splits among these groups, and hierarchies among and within them. But all their members are people with specialized training and knowledge, and all of them enjoy greater prestige and income than their immediate (and their distant) predecessors in Europe and America. They are the "clerks" of the modern world. Are they a "new class," sharing interests and consciousness? Are they the long-awaited *next* class?

From the time of Marx until the present day, the most common goal of social research has been to discover the next ruling class. Virtually no one believed or believes in the staying power of the bourgeoisie. Surely there has never been a less prepossessing group of political leaders than the merchants and industrialists of the nineteenth and twentieth centuries. Aware of their own limitations, they have often preferred to leave office to old aristocrats, who have the style for it, or to new professionals, who have the wits. Since capital is a potent force, aristocrats and professionals mostly do its bidding. When they attempt defiance, they look for support among the people, and it is a special contingent from among the people —industrial workers, steeled by class struggle—who have commonly been described as the legitimate heirs of the bourgeoisie. Marx's argument focuses not on the workers' right to succeed but on their will and capacity to do so. He thought that their numbers and then their solidarity would do for them what wealth had done for merchants and industrialists. They might need intellectual support; so had the bourgeoisie. But only the workers, by virtue of their economic position and social strength, could constitute an alternative class.

But the workers have not taken power and held it—anywhere—and in the last three or four decades a number of writers, like James Burnham and Milovan Djilas, have begun to make the case for an alternative succession of intellectuals and managers. Now the Hungarian writers George Konrád and Ivan Szelényi, and the American writer Alvin Gouldner, make the case again; Gouldner with rather more enthusiasm than seems warranted by anything we currently know, Konrád and Szelényi with a new theoretical precision and density. Their argument is best understood as a critique of Marx. But this is a critique worked out (by Burnham and Djilas too) within what can still be called the Marxist tradition of class analysis. I am not sure whether it is a sign of the strength of that tradition, or of its disintegration, that it can now accommodate predictions so radically divergent from Marx's own.

Both the accommodation and the divergence are apparent in the way these writers use Marxism against itself. They are sociological Marxists who treat political Marxism, and socialism more generally, as the ideology of the new class. The argument is clearest in the case of Eastern Europe—and therefore in the version of Konrád and Szelényi. Landlords and capitalists were expropriated in the name of the workers, they argue, but the chief agents of socialization and its chief beneficiaries were and are the intellectuals (professional revolutionaries and the technical experts they recruit), who now control the economic surplus, "maximize" it through central planning or simple confiscation, and distribute it on scientific or ideological principles. They replace the market with the plan, the economy with the state, the old ruling class with themselves.

Konrád and Szelényi's essay is a piece of *samizdat,* a banned book in Hungary, written in 1974, smuggled out of the country, and finally published in the West only last year. Its authors insist that they do not intend a critique or a justification of the new class but only an "imminent structural analysis." Their book is, nevertheless, brutally deflationary because it treats the intelligentsia in exactly the same way that classical Marxism treats the bourgeoisie: the new class is *only* the next class, not (like the Marxist proletariat) the last class. Its members pursue particular interests; they don't advance universal values. Of course, the vanguard of the new class (the Communist party) claims to advance universal values, just as bourgeois vanguards (Puritans and Jacobins) did in their own revolutionary period. But vanguard rule, now as then, culminates in dictatorship and terror and eventually produces a reaction by the new class, whose members demand their own version of normalcy. The vanguard/class tension is today represented by the struggles, in countries like Czechoslovakia, Hungary, and Poland, between the party and the technical intelligentsia. "Socialism with a human face" is the creed of the intelligentsia, and if we ever get a close look at that face, Konrád and Szelényi suggest, we will see that it is not human merely, it is the face of particular men and women, economists, technocrats, managers, and professionals. The technocratic Thermidor is, or will be, more humane and liberal than the party's Terror, but it is still a form of class rule, not the rule of the associated workers, where everyone is a worker, or of the people generally.

Gouldner's view of the new class is considerably more optimistic. Indeed, his book is written with a kind of buoyancy, only occasionally curbed, as if he himself were marching along the road to class power, sometimes beset by anxieties, sometimes bursting into song. "The New Class is the most progressive force in modern society . . . a center of whatever human emancipation is possible in the foreseeable future." And again, "The New Class is the universal class in embryo but badly flawed." I don't know if that last sentence means that the contemporary intelligentsia represents, so to speak, the infancy of universality, or if it means that the infant is going to grow up deformed. The first meaning is more likely. For Gouldner does not view the new class as a group of men and women defined by their social position, using state power, as the bourgeoisie used capital, to control and distribute the economic surplus. Instead, the new class is defined in terms of its culture; it is a "speech community." This is not the same thing as a linguistic community: intellectuals are a new class, thank heaven, not a new nation. What they share is the "culture of critical discourse"—CCD in Gouldner's text, where he continually writes as if the new class has CCD in much the same way that capitalists have capital, homeowners have homes, and cancer victims have cancer. At the same time, he insists that critical discourse is by its very nature no one's exclusive possession: it is open, available, anti-authoritarian, consensual . . . universal. As a ruling class, intellectuals have to give reasons; they can't just give orders.

But it is absurdly easy to give reasons for giving orders, and I am not sure that the orders are any different, at the receiving end, if they come with, or if they come without, explanations. In any case, there is a sleight of hand involved in treating the capacity for critical discourse as the "capital" of the new class. This formulation leads Gouldner to deny membership to state and corporate bureaucrats, since their culture, he argues, is one of routinized obedience, hierarchy, and official secrecy. In contrast to Hegel, who thought the civil service the very model of a universal class (because its speciality, so to speak, is the general welfare), Gouldner insists that bureaucrats merely follow orders; they are the agents, dull and grey, of other social groups. But when he wants to impress us with the recent growth and current size of the new class, bureaucrats and managers are included in its ranks.[1] And indeed, without "line officials" in government and corporations, the new class is not very impressive; nor is it easy to see how it might one day rule over the rest of us. If there is a new class, bureaucracy is its cutting edge. One can't construct sociological categories by excluding groups one finds unlovely, or uncritical.

By contrast, Konrád and Szelényi provide a more inclusive description. The new class, for them, consists of "three partners of equal importance"—the stratum of economists and technocrats, the administrative and political bureaucracy, and the ideological, scientific, and artistic intelligentsia. The last group is somewhat problematic (it includes "marginal intellectuals" like themselves), and I will come back to it later. What is crucial is that all or almost all these people share perceptions and interests: they are "mutually dependent on one

[1] See the table on p. 15 of *The Future of Intellectuals*.

another and impregnated with one another's logic." It is their standing vis-à-vis the state and their common sense of the purpose of the state that constitute their class unity. The conditions that produce this unity are, however, historically specific. Konrád and Szelényi's account of that specificity is probably the most brilliant part of their book, a sustained and powerful analysis of East European social structure since the Middle Ages. The role of the central state, the prestige of office-holding, the weakness of the bourgeoisie and its inability to support an "organic intelligentsia" of its own—all this helps explain the appearance of an "intelligentsia of office" replacing the bureaucratic gentry and nobility of an earlier time.

It's not the case in Eastern Europe that the new class replaces the bourgeoisie, and I am inclined to doubt Gouldner's claim that it is in the process of doing so in the West. What is actually happening is rather different, and is best approached, I think, by way of an analogy. Just as a parallel nobility, the *noblesse de robe*, established itself during the early modern period of state centralization, so a parallel bourgeoisie, which we might call the *bourgeoisie de robe*, is in the process of establishing itself today. Like the absolutist state of the seventeenth century, so the planning and welfare state of our own time, and the giant corporations that exist alongside it, provide new opportunities for social advance. For a number of reasons, however, it is a mistake to describe the men and women who take advantage of these opportunities as the members of a new class. First, the *bourgeoisie de robe* reproduces the division of state and economy that is the most fundamental characteristic of bourgeois life. It isn't, that is, exclusively associated with state power; it develops also within the corporate world and retains a firm commitment to private property. And second, its life style is individualist and consumerist in the conventional bourgeois fashion. The claim that there are new ways of life, like the claim that there is a "new politics," associated with the new class hardly bears examination. Gouldner writes, for example, that the members were united "in their opposition to the United States' war on Vietnam." But that's only true, again, if one excludes from membership all those people—engineers, managers, bureaucrats—who were not significant participants in the antiwar movement. Then the "new class" also fits nicely into neoconservative fantasies, but it's not an actual social formation. We might argue instead that neoconservatism, with its simultaneous advocacy of nonideological planning and market economics, most accurately represents the consciousness of the social groups that Gouldner describes. And so, it makes sense to call them a neobourgeoisie, whose members owe their middle and upper middle-class status to their ability to earn educational certificates and to occupy a range of positions in state and economy.

The integration of new intelligentsia and old bourgeoisie is a distinguishing feature of Western social structure. It has been worked out with very little dislocation and without any (visible) form of vanguard politics. The old rulers of our state and economy have not been displaced; they have been joined. And since the bourgeoisie was never an exclusive club, the joining was easy. It certainly did not entail a massive shift in power relations. With whatever qualifications, it still makes sense to describe countries like the United States as capitalist societies.

But that is not by any means the whole story of East/West differences. For Western political structures also have a distinguishing feature, stressed by Konrád and Szelényi, though not by Gouldner: "the sovereignty of a political mechanism based on representation." Wealth and technical knowledge are two forms of legitimacy that are in competition, wherever democratic institutions exist, with political representation. Perhaps that's only to say that we are ruled by politicians as well as by capitalists and technocrats. Given the current reputation of politicians, that may not mean much. In fact, however, the principle of representation opens large and significant possibilities—first, because it guarantees that politics is not a matter merely of purchase or expertise but a competition of would-be representatives, and second, because it gives a voice or a potential voice to all those men and women without either wealth or technical knowledge upon whose suffrage politicians must depend. It also brings intellectuals into political life—as spokesmen, publicists, and agitators. They defend interests, invent and criticize positions. The ideological conflicts of a democratic society are impossible without them, and while it can be said that they always pursue their own secret goals, it does not appear that they ever agree on what those goals are. Endlessly divided, they make connections with different and opposing social forces—so long as there are different and opposing forces in the political arena.

But what happens when the other forces are weak and the intellectuals themselves—or rather, the intelligentsia as a whole, bureaucrats, technocrats, and so on—come to power? Konrád and Szelényi worry that they are then unable to articulate in critical fashion either their own or anyone else's interests and ideologies. They are the prisoners of a collective mythology, silent and acquiescent in the face of their own pretensions—unable, for example, to mount a serious campaign against Leninist vanguard theory. And so the denial of representation and the closing down of ideological conflict in Eastern Europe are only the political recognition of a kind of epistemological closure. Even were conflict open and free, intellectuals would do nothing more than repeat the official line, for they are themselves (at last!) the officials. But the book that Konrád and Szelényi have written, though it can't be published in the country where they wrote it, is significant evidence against this argument. These two are *marginal* intellectuals; they still participate in the culture of criticism from which their class associates, the intelligentsia of office, have evidently absented themselves. And it is certain that the liberalization of political life in Hungary would reveal others like themselves, ready to cut loose from the new class, to speak for other groups, to challenge state power, and even to defend political liberalism and economic self-management: the mirror image of their own intellectual freedom. And in the United States, where conflict is open and free, the status, interests, and ideology of the neobourgeoisie are already the subject of critical scrutiny by both friendly and hostile intellectuals, none of whom are simply members (though a few aspire to be nothing more).

Perhaps we should say straightforwardly that marginal intellectuals are the only real intellectuals: Konrád and Szelényi are prime examples. They themselves deny the equation and make fun of Sartre's remark that an atomic physicist is an intellectual only when he signs a petition against nuclear testing. No doubt, that is a piece of sociological, perhaps also of political, silliness. So is Gouldner's attempt to define the intelligentsia as a class by reference to its critical culture. Still, there is a point here. Like Shelley's poets, intellectuals belong to a category that isn't only sociological but also and more importantly normative. The intelligentsia of office and the *bourgeoisie de robe*: these are sociological categories, and we assign members to them on the basis of their social position. But we recognize intellectuals by other marks. They are committed to rigorous analysis, "imminent critique," truth-telling. Or better, they are poets too (Konrád is a brilliant novelist), legislators for the mind and spirit. And because of that, they can never be, we can never conceive of them to be, the members of a ruling class. They always have an interest in the class that comes next.

(1980)

PART IV

Democratic Socialism

15

In Defense of Equality

I

At the very center of conservative thought lies this idea: that the present division of wealth and power corresponds to some deeper reality of human life. Conservatives don't want to say merely that the present division is what it ought to be, for that would invite a search for some distributive principle—as if it were possible to *make* a distribution. They want to say that whatever the division of wealth and power is, it naturally is, and that all efforts to change it, temporarily successful in proportion to their bloodiness, must be futile in the end. We are then invited, as in Irving Kristol's recent *Commentary* article, to reflect upon the perversity of those who would make the attempt.[1] Like a certain sort of leftist thought, conservative argument seems quickly to shape itself around a rhetoric of motives rather than one of reasons. Kristol is especially adept at that rhetoric and strangely unconcerned about the reduc-

[1] "About Equality," *Commentary*, November 1972.

237

tionism it involves. He aims to expose egalitarianism as the ideology of envious and resentful intellectuals. No one else cares about it, he says, except the "new class" of college-educated, professional, most importantly, professorial men and women, who hate their bourgeois past (and present) and long for a world of their own making.

I suppose I should have felt, after reading Kristol's piece, that the decent drapery of my socialist convictions has been stripped away, that I was left naked and shivering, small-minded and self-concerned. Perhaps I did feel a little like that, for my first impulse was to respond in kind, exposing anti-egalitarianism as the ideology of those other intellectuals—"they are mostly professors, of course"—whose spiritual path was sketched some years ago by the editor of *Commentary*. But that would be at best a degrading business, and I doubt that my analysis would be any more accurate than Kristol's. It is better to ignore the motives of these "new men" and focus instead on what they say: that the inequalities we are all familiar with are inherent in our condition, are accepted by ordinary people (like themselves), and are criticized only by the perverse. I think all these assertions are false; I shall try to respond to them in a serious way.

Kristol doesn't argue that we can't possibly have greater equality or greater inequality than we presently have. Both communist and aristocratic societies are possible, he writes, under conditions of political repression or economic under-development and stagnation. But insofar as men and women are set free from the coerciveness of the state and from material necessity, they will distribute themselves in a more natural way, more or less as contemporary Americans have done. The American way is exemplary because it derives from or reflects the real inequalities of mankind. People don't naturally fall into two classes (patricians and plebeians) as conservatives once thought; nor can they plausibly be grouped into a single class (citizens or comrades) as leftists still believe. Instead, "human talents and abilities . . . distribute themselves along a bell-shaped curve, with most people clustered around the middle, and with much smaller percentages at the lower and higher ends." The marvels of social science!—this distribution is a demonstrable fact. And it is another "demonstrable fact that in all modern bourgeois societies, the distribution of income is also along a bell-shaped curve. . . ." The second bell echoes the first. Moreover, once this harmony is established, "the political structure—the distribution of political power— follows along the same way. . . ." At this point, Kristol must add, "however slowly and reluctantly," since he believes that the Soviet economy is moving closer every year to its natural shape, and it is admittedly hard to find evidence that nature is winning out in the political realm. But in the United States, nature is triumphant: we are perfectly bell-shaped.

The first bell is obviously the crucial one. The defense of inequality reduces to these two propositions: that talent is distributed unequally and that talent will out. Clearly, we all want men and women to develop and express their talents, but whenever they are able to do that, Kristol suggests, the bell-shaped curve will appear or reappear, first in the economy, then in the political system. It is a neat argument but also a peculiar one, for there is no reason to think that "human talents and abilities" in fact distribute themselves along a *single* curve, although income necessarily does. Consider the range and variety of human capacities: intelligence, physical strength, agility and grace, artistic creativity, mechanical skill, leadership, endurance, memory, psychological insight, the capacity for hard work—even, moral strength, sensitivity, the ability to express compassion. Let's assume that with respect to all these, most people (but different people in each case) cluster around the middle of whatever scale we can construct, with smaller numbers at the lower and higher ends. Which of these curves is actually echoed by the income bell? Which, if any, ought to be?

There is another talent that we need to consider: the ability to make money, the green thumb of bourgeois society—a secondary talent, no doubt, combining many of the others in ways specified by the immediate environment, but probably also a talent that distributes, if we could graph it, along a bell-shaped curve. Even this curve would not correlate exactly with the income bell because of the intervention of luck, that eternal friend of the untalented, whose most important social expression is the inheritance of property. But the correlation would be close enough, and it might also be morally plausible and satisfying. People who are able to make money ought to make money, in the same way that people who are able to write books ought to write books. Every human talent should be developed and expressed.

The difficulty here is that making money is only rarely a form of self-expression, and the money we make is rarely enjoyed for its intrinsic qualities (at least, economists frown upon that sort of enjoyment). In a capitalist world, money is the universal medium of exchange; it enables the men and women who possess it to purchase virtually every other sort of social good; we collect it for its exchange value. Political power, celebrity, admiration, leisure, works of art, baseball teams, legal advice, sexual pleasure, travel, education, medical care, rare books, sailboats—all these (and much more) are up for sale. The list is as endless as human desire and social invention. Now isn't it odd, and morally implausible and unsatisfying, that all these things should be distributed to people with a talent for making money? And even odder and more unsatisfying that they should be distributed (as they are) to people who have money, whether or not they made it, whether or not they possess any talent at all?

Rich people, of course, always look talented—just as the beautiful people always look beautiful—to the deferential observer. But it is the first task of social science, one would think, to look beyond these appearances. "The properties of money," Marx wrote, "are my own (the possessor's) properties and faculties. What I *am* and *can do* is, therefore, not at all determined by my individuality. I *am* ugly, but I can buy the most beautiful woman for myself. Consequently, I am not ugly, for the effect of ugliness, its power to repel, is annulled by money. . . . I am a detestable, dishonorable, unscrupulous, and stupid man, but money is honored and so also is its possessor."[2]

It would not be any better if we gave people money in direct proportion to their intelligence, their strength, or their moral rectitude. The resulting distributions would each, no doubt, reflect what Kristol calls "the tyranny of the bell-shaped curve," though it is worth noticing again that the populations in the lower, middle, and upper regions of each graph would be radically different. But whether it was the smart, the strong, or the righteous who enjoyed all the things that money can buy, the oddity would remain: why them? Why anybody? In fact, there is no single talent or combination of talents that plausibly entitles a man to every available social good—and there is no single talent or combination of talents that necessarily must win the available goods of a free society. Kristol's bell-shaped curve is tyrannical only in a purely formal sense. Any particular distribution may indeed be bell-shaped, but there are a large number of possible distributions. Nor need there be a single distribution of all social goods, for different goods might well be distributed differently. Nor again need all these distributions follow this or that talent curve, for in the sharing of some social goods, talent does not seem a relevant consideration at all.

Consider the case of medical care: surely it should not be distributed to individuals because they are wealthy, intelligent, or righteous, but only because they are sick. Now, over any given period of time, it may be true that some men and women won't require any medical treatment, a very large number will need some moderate degree of attention, and a few will have to have intensive care. If that is so, then we must hope for the appearance of another bell-shaped curve.

[2] *Early Writings*, trans. T. B. Bottomore (London, 1963), p. 191.

Not just any bell will do. It must be the right one, echoing what might be called the susceptibility-to-sickness curve. But in America today, the distribution of medical care actually follows closely the lines of the income graph. It's not how a man feels, but how much money he has that determines how often he visits a doctor. Another demonstrable fact! Does it require envious intellectuals to see that something is wrong?

There are two possible ways of setting things right. We might distribute income in proportion to susceptibility-to-sickness, or we might make sure that medical care is not for sale at all, but is available to those who need it. The second of these is obviously the simpler. Indeed, it is a modest proposal and already has wide support, even among those ordinary men and women who are said to be indifferent to equality. And yet, the distribution of medical care solely for medical reasons would point the way toward an egalitarian society, for it would call the dominance of the income curve dramatically into question.

II

What egalitarianism requires is that many bells should ring. Different goods should be distributed to different people for different reasons. Equality is not a simple notion, and it cannot be satisfied by a single distributive scheme—not even, I hasten to add, by a scheme that emphasizes need. "From each according to his abilities, to each according to his needs" is a fine slogan with regard to medical care. Tax money collected from all of us in proportion to our resources (these will never correlate exactly with our abilities, but that problem I shall leave aside for now) must pay the doctors who care for those of us who are sick. Other people who deliver similar sorts of social goods should probably be paid in the same way —teachers and lawyers, for example. But Marx's slogan doesn't help at all with regard to the distribution of political power, honor and fame, leisure time, rare books, and sailboats. None of these things can be distributed to individuals in proportion to their needs, for they are not things that anyone (strictly speaking) needs. They can't be distributed in equal amounts or given to whoever wants them, for some of them are necessarily scarce, and some of them can't be possessed unless other people agree on the proper name of the possessor. There is no criterion, I think, that will fit them all. In the past they have indeed been distributed on a single principle: men and women have possessed them or their historical equivalents because they were strong or well-born or wealthy. But this only suggests that a society in which any single distributive principle is dominant cannot be an egalitarian society. Equality requires a diversity of principles, which mirrors the diversity both of mankind and of social goods.

Whenever equality in this sense does not prevail, we have a kind of tyranny, for it is tyrannical of the well-born or the strong or the rich to gather to themselves social goods that have nothing to do with their personal qualities. This is an idea beautifully expressed in a passage from Pascal's *Pensées*, which I am going to quote at some length, since it is the source of my own argument.[3]

The nature of tyranny is to desire power over the whole world and outside its own sphere.

There are different companies—the strong, the handsome, the intelligent, the devout—and each man reigns in his own, not elsewhere. But sometimes they meet, and the strong and the handsome fight for mastery—foolishly, for their mastery is of different kinds. They misunderstand one another, and make

[3] I am also greatly indebted to Bernard Williams, in whose essay "The Idea of Equality" (first published in Laslett and Runciman, *Philosophy, Politics and Society*, second series [Oxford, 1962]) a similar argument is worked out. The example of medical care, to which I recur, is suggested by him. The Pascal quote is from J. M. Cohen's translation of *The Pensées* (London and Baltimore, 1961). no. 244.

the mistake of each aiming at universal dominion. Nothing can win this, not even strength, for it is powerless in the kingdom of the wise. . . .

Tyranny. The following statements, therefore, are false and tyrannical: "Because I am handsome, so I should command respect." "I am strong, therefore men should love me. . . ." "I am . . . etc."

Tyranny is the wish to obtain by one means what can only be had by another. We owe different duties to different qualities: love is the proper response to charm, fear to strength, and belief to learning.

Marx makes a very similar argument in one of the early manuscripts; perhaps he had this *pensée* in mind.[4]

Let us assume man to be man, and his relation to the world to be a human one. Then love can only be exchanged for love, trust for trust, etc. If you wish to enjoy art you must be an artistically cultivated person; if you wish to influence other people, you must be a person who really has a stimulating and encouraging effect upon others. . . . If you love without evoking love in return, i.e., if you are not able, by the manifestation of yourself as a loving person, to make yourself a beloved person, then your love is impotent and a misfortune.

The doctrine suggested by these passages is not an easy one, and I can expound it only in a tentative way. It isn't that every man should get what he deserves—as in the old definition of justice—for desert is relevant only to some of the exchanges that Pascal and Marx have in mind. Charming men and women don't deserve to be loved: I may love this one or that one, but it can't be the case that I ought to do so. Similarly, learned men don't deserve to be believed: they are believed or not depending on the arguments they make. What Pascal and Marx

---

[4] *Early Writings,* pp. 193–94.

are saying is that love and belief can't rightly be had in any other way—can't be purchased or coerced, for example. It is wrong to seek them in any way that is alien to their intrinsic character. In its extended form, their argument is that for all our personal and collective resources, there are distributive reasons that are somehow *right*, that are naturally part of our ideas about the things themselves. So nature is re-established as a critical standard, and we are invited to wonder at the strangeness of the existing order.

This new standard is egalitarian, even though it obviously does not require an equal distribution of love and belief. The doctrine of right reasons suggests that we pay equal attention to the "different qualities," and to the "individuality" of every man and woman, that we find ways of sharing our resources that match the variety of their needs, interests, and capacities. The clues that we must follow lie in the conceptions we already have, in the things we already know about love and belief, and also about respect, obedience, education, medical care, legal aid, all the necessities of life—for this is no esoteric doctrine, whatever difficulties it involves. Nor is it a panacea for human misfortune, as Marx's last sentence makes clear: it is only meant to suggest a humane form of social accommodation. There is little we can do, in the best of societies, for the man who isn't loved. But there may be ways to avoid the triumph of the man who doesn't love—who buys love or forces it—or at least of his parallels in the larger social and political world: the leaders, for example, who are obeyed because of their coercive might or their enormous wealth. Our goal should be an end to tyranny, a society in which no human being is master outside his sphere. That is the only society of equals worth having.

But it isn't readily had, for there is no necessity implied by the doctrine of right reasons. Pascal is wrong to say that "strength is powerless in the kingdom of the wise"—or rather, he is talking of an ideal realm and not of the intellectual world as we know it. In fact, wise men and women (at any rate, smart men and women) have often in the past defended the tyranny of the strong, as they still defend the tyranny of the rich. Sometimes, of course, they do this because they are persuaded of the necessity or the utility of tyrannical rule; sometimes for other reasons. Kristol suggests that whenever intellectuals are not persuaded, they are secretly aspiring to a tyranny of their own: they too would like to rule outside their sphere. Again, that's certainly true of some of them, and we all have our own lists. But it's not necessarily true. Surely it is possible, though no doubt difficult, for an intellectual to pay proper respect to the "different companies." I want to argue that in our society the only way to do that, or to begin to do it, is to worry about the tyranny of money.

III

Let's start with some things that money cannot buy. It can't buy the American League pennant: star players can be hired, but victories presumably are not up for sale. It can't buy the National Book Award: writers can be subsidized, but the judges presumably can't be bribed. Nor, it should be added, can the pennant or the award be won by being strong, charming, or ideologically correct—at least we all hope not. In these sorts of cases, the right reasons for winning are built into the very structure of the competition. I am inclined to think that they are similarly built into a large number of social practices and institutions. It's worth focusing again, for example, on the practice of medicine. From ancient times, doctors were required to take an oath to help the sick, not the powerful or the wealthy. That requirement reflects a common understanding about the very nature of medical care. Many professionals don't share that understanding, but the opinion of ordinary men and women, in this case at least, is profoundly egalitarian.

The same understanding is reflected in our legal system. A

man accused of a crime is entitled to a fair trial simply by virtue of being an accused man; nothing else about him is a relevant consideration. That is why defendants who cannot afford a lawyer are provided with legal counsel by the state: otherwise justice would be up for sale. And that is why defense counsel can challenge particular jurors thought to be prejudiced: the fate of the accused must hang on his guilt or innocence, not on his political opinions, his social class, or his race. We want different defendants to be treated differently, but only for the right reasons.

The case is the same in the political system, whenever the state is a democracy. Each citizen is entitled to one vote simply because he is a citizen. Men and women who are ambitious to exercise greater power must collect votes, but they can't do that by purchasing them; we don't want votes to be traded in the marketplace, though virtually everything else is traded there, and so we have made it a criminal offense to offer bribes to voters. The only right way to collect votes is to campaign for them, that is, to be persuasive, stimulating, encouraging, and so on. Great inequalities in political power are acceptable only if they result from a political process of a certain kind, open to argument, closed to bribery and coercion. The freely given support of one's fellow citizens is the appropriate criterion for exercising political power and, once again, it is not enough, or it shouldn't be, to be physically powerful, or well-born, or even ideologically correct.

It is often enough, however, to be rich. No one can doubt the mastery of the wealthy in the spheres of medicine, justice, and political power, even though these are not their own spheres. I don't want to say, their unchallenged mastery, for in democratic states we have at least made a start toward restricting the tyranny of money. But we have only made a start: think how different America would have to be before these three companies of men and women—the sick, the accused, the politically ambitious—could be treated in strict accordance with their individual qualities. It would be immediately necessary to have a national health service, national legal assistance, the strictest possible control over campaign contributions. Modest proposals, again, but they represent so many moves toward the realization of that old socialist slogan about the abolition of money. I have always been puzzled by that slogan, for socialists have never, to my knowledge, advocated a return to a barter economy. But it makes a great deal of sense if it is interpreted to mean *the abolition of the power of money outside its sphere*. What socialists want is a society in which wealth is no longer convertible into social goods with which it has no intrinsic connection.

But it is in the very nature of money to be convertible (that's all it is), and I find it hard to imagine the sorts of laws and law enforcement that would be necessary to prevent monied men and women from buying medical care and legal aid over and above whatever social minimum is provided for everyone. In the United States today, people can even buy police protection beyond what the state provides, though one would think that it is the primary purpose of the state to guarantee equal security to all its citizens, and it is by no means the rich, despite the temptations they offer, who stand in greatest need of protection. But this sort of thing could be prevented only by a very considerable restriction of individual liberty—of the freedom to offer services and to purchase them. The case is even harder with respect to politics itself. One can stop overt bribery, limit the size of campaign contributions, require publicity, and so on. But none of these things will be enough to prevent the wealthy from exercising power in all sorts of ways to which their fellow citizens have never consented. Indeed, the ability to hold or spend vast sums of money is itself a form of power, permitting what might be called preemptive strikes against the political system. And this, it seems to me, is the strongest possible argument for a radical redistribution of wealth. So long as money is convertible outside its sphere, it must be widely and more or less equally held so as to minimize its distorting effects upon legitimate distributive processes.

IV

What is the proper sphere of wealth? What sorts of things are rightly had in exchange for money? The obvious answer is also the right one: all those economic goods and services, beyond what is necessary to life itself, that men and women find useful or pleasing. There is nothing degraded about wanting these things; there is nothing unattractive, boring, debased, or philistine about a society organized to provide them for its members. Kristol insists that a snobbish dislike for the sheer productivity of bourgeois society is a feature of egalitarian argument. I would have thought that a deep appreciation of that productivity has more often marked the work of socialist writers. The question is, how are the products to be distributed? Now, the right way to possess useful and pleasing things is by making them, or growing them, or somehow providing them for others. The medium of exchange is money, and this is the proper function of money and, ideally, its only function.

There should be no way of acquiring rare books and sailboats except by working for them. But this is not to say that workers deserve whatever money they can get for the goods and services they provide. In capitalist society, the actual exchange value of the work they do is largely a function of market conditions over which they exercise no control. It has little to do with the intrinsic value of the work or with the individual qualities of the worker. There is no reason for socialists to respect it, unless it turns out to be socially useful to do so. There are other values, however, that they must respect, for money isn't the only or necessarily the most important thing for which work can be exchanged. A lawyer is surely entitled to the respect he wins from his colleagues and to the gratitude and praise he wins from his clients. The work he has done may also constitute a good reason for making him director of the local legal aid society; it may even be a good reason for making him a ju lge. It isn't, on the face of it, a good reason for allow-ing him an enormous income. Nor is the willingness of his clients to pay his fees a sufficient reason, for most of them almost certainly think they should be paying less. The money they pay is different from the praise they give, in that the first is extrinsically determined, the second freely offered.

In a long and thoughtful discussion of egalitarianism in the *Public Interest*, Daniel Bell worries that socialists today are aiming at an "equality of results" instead of the "just meritocracy" (the career open to talents) that he believes was once the goal of leftist and even of revolutionary politics.[5] I confess that I am tempted by "equality of results" in the sphere of money, precisely because it is so hard to see how a person can merit the things that money can buy. On the other hand, it is easy to list cases where merit (of one sort or another) is clearly the right distributive criteria, and where socialism would not require the introduction of any other principle.

· Six people speak at a meeting, advocating different policies, seeking to influence the decision of the assembled group.
· Six doctors are known to aspire to a hospital directorship.
· Six writers publish novels and anxiously await the reviews of the critics.
· Six men seek the company and love of the same woman.

Now, we all know the right reasons for the sorts of decisions, choices, judgments that are in question here. I have never heard anyone seriously argue that the woman must let herself be shared, or the hospital establish a six-man directorate, or the critics distribute their praise evenly, or the people at the meeting adopt all six proposals. In all these cases, the personal qualities of the individuals involved, or the arguments they make, or the work they do (as these appear to the others) should carry the day.

But what sorts of personal qualities are relevant to owning

[5] "On Meritocracy and Equality," *Public Interest*, Fall 1972.

a $20,000 sailboat? A love for sailing, perhaps, and a willingness to build the boat or to do an equivalent amount of work. In America today, it would take a steelworker about two years to earn that money (assuming that he didn't buy anything else during all that time) and it would take a corporation executive a month or two. How can that be right, when the executive also has a rug on the floor, air-conditioning, a deferential secretary, and enormous personal power? He is being paid as he goes, while the steelworker is piling up a kind of moral merit (so we have always been taught) by deferring pleasure. Surely there is no meritocratic defense for this sort of difference. It would seem much better to pay the worker and the executive more or less the same weekly wage and let the sailboat be bought by the person who is willing to forgo other goods and services, that is, by the person who really wants it. Is this "equality of result"? In fact, the results will be different, if the people are, and it seems to me that they will be different for the right reasons.

Against this view, there is a conventional but also very strong argument that can be made on behalf of enterprise and inventiveness. If there is a popular defense of inequality, it is this one, but I don't think it can carry us very far toward the inequalities that Kristol wants to defend. Consider the case of the man who builds a better mousetrap, or opens a restaurant and sells delicious blintzes, or does a little teaching on the side. He has no air-conditioning, no secretary, no power; probably his reward has to be monetary. He has to have a chance, at least, to earn a little more money than his less enterprising neighbors. The market doesn't guarantee that he will in fact earn more, but it does make it possible, and until some other way can be found to do that, market relations are probably defensible under the doctrine of right reasons. Here in the world of the petty-bourgeoisie, it seems appropriate that people able to provide goods or services that are novel, timely, or particularly excellent should reap the rewards they presumably had in mind when they went to work. And that they were right to have in mind: no one would want to feed blintzes to strangers, day after day, merely to win their gratitude.

But one might well want to be a corporation executive, day after day, merely to make all those decisions. It is precisely the people who are paid or who pay themselves vast sums of money who reap all sorts of other rewards too. We need to sort out these different forms of payment. First of all, there are rewards, like the pleasure of exercising power, that are intrinsic to certain jobs. An executive must make decisions— that's what he is there for—and even decisions seriously affecting other people. It is right that he should do that, however, only if he has been persuasive, stimulating, encouraging, and so on, and won the support of a majority of those same people. That he owns the corporation or has been chosen by the owners isn't enough. Indeed, given the nature of corporate power in contemporary society, the following statement (to paraphrase Pascal) is false and tyrannical: because I am rich, so I should make decisions and command obedience. Even in corporations organized democratically, of course, the personal exercise of power will persist. It is more likely to be seen, however, as it is normally seen in political life—as the chief attraction of executive positions. And this will cast a new light on the other rewards of leadership.

The second of these consists in all the side-effects of power: prestige, status, deference, and so on. Democracy tends to reduce these, or should tend that way when it is working well, without significantly reducing the attractions of decision making. The same is true of the third form of reward, money itself, which is owed to work, but not necessarily to place and power. We pay political leaders much less than corporation executives, precisely because we understand so well the excitement and appeal of political office. Insofar as we recognize the political character of corporations, then, we can pay their executives less too. I doubt that there would be a lack of candidates even if we paid them no more than was paid to any other corporation employee. Perhaps there are reasons for paying them more—but not meritocratic reasons, for we give all the attention that is due to their merit when we make them our leaders.

We do not give all due attention to the restaurant owner, however, merely by eating his blintzes. Him we have to pay, and he can ask, I suppose, whatever the market will bear. That's fair enough, and no real threat to equality so long as he can't amass so much money that he becomes a threat to the integrity of the political system and so long as he does not exercise power, tyrannically, over other men and women. Within his proper sphere, he is as good a citizen as any other. His activities recall Dr. Johnson's remark: "There are few ways in which man can be more innocently employed than in getting money."

V

The most immediate occasion of the conservative attack on equality is the reappearance of the quota system—newly designed, or so it is said, to move us closer to egalitarianism rather than to maintain old patterns of religious and racial discrimination. Kristol does not discuss quotas, perhaps because they are not widely supported by professional people (or by professors): the disputes of the last several years do not fit the brazen simplicity of his argument. But almost everyone else talks about them, and Bell worries at some length, and rightly, about the challenge quotas represent to the "just meritocracy" he favors. Indeed, quotas in any form, new or old, establish "wrong reasons" as the basis of important social decisions, perhaps the most important social decisions: who shall be a doctor, who shall be a lawyer, and who shall be a bureaucrat. It is obvious that being black or a woman or having a Spanish surname (any more than being white, male, and Protestant) is no qualification for entering a university or a medical school or joining the civil service. In a sense, then, the critique of quotas consists almost entirely of a series of restatements and reiterations of the argument I have been urging in this essay. One only wishes that the critics would apply it more generally than they seem ready to do. There is more to be said, however, if they consistently refuse to do that.

The positions for which quotas are being urged are, in America today, key entry points to the good life. They open the way, that is, to a life marked above all by a profusion of goods, material and moral: possessions, conveniences, prestige, and deference. Many of these goods are not in any plausible sense appropriate rewards for the work that is being done. They are merely the rewards that upper classes throughout history have been able to seize and hold for their members. Quotas, as they are currently being used, are a way of redistributing these rewards by redistributing the social places to which they conventionally pertain. It is a bad way, because one really wants doctors and (even) civil servants to have certain sorts of qualifications. To the people on the receiving end of medical and bureaucratic services, race and class are a great deal less important than knowledge, competence, courtesy, and so on. I don't want to say that race and class are entirely unimportant: it would be wrong to underestimate the distortions introduced by an inegalitarian society into these sorts of human relations. But if the right reason for receiving medical care is being sick, then the right reason for giving medical care is being able to help the sick. And so medical schools should pay attention, first of all and almost exclusively, to the potential helpfulness of their applicants.

But they may be able to do that only if the usual connections between place and reward are decisively broken. Here is another example of the doctrine of right reasons. If men and women wanted to be doctors primarily because they wanted to be helpful, they would have no reason to object when judgments were made about their potential helpfulness. But so long as there are extrinsic reasons for wanting to be a doctor, there will be pressure to choose doctors (that is, to make medical school places available) for reasons that are similarly extrinsic.

So long as the goods that medical schools distribute include more than certificates of competence, include, to be precise, certificates of earning power, quotas are not entirely implausible. I don't see that being black is a worse reason for owning a sailboat than being a doctor. They are equally bad reasons.

Quotas today are a means of lower-class aggrandizement, and they are likely to be resolutely opposed, opposed without guilt and worry, only by people who are entirely content with the class structure as it is and with the present distribution of goods and services. For those of us who are not content, anxiety cannot be avoided. We know that quotas are wrong, but we also know that the present distribution of wealth makes no moral sense, that the dominance of the income curve plays havoc with legitimate distributive principles, and that quotas are a form of redress no more irrational than the world within which and because of which they are demanded. In an egalitarian society, however, quotas would be unnecessary and inexcusable.

VI

I have put forward a difficult argument in very brief form, in order to answer Kristol's even briefer argument—for he is chiefly concerned with the motives of those who advocate equality and not with the case they make or try to make. He is also concerned, he says, with the fact that equality has suddenly been discovered and is now for the first time being advocated as the *chief* virtue of social institutions: as if societies were not complex and values ambiguous. I don't know what discoverers and advocates he has in mind.[6] But it

---

[6] The only writer he mentions is John Rawls, whose *Theory of Justice* (Cambridge, Mass., 1971) Kristol seems entirely to misunderstand. For Rawls explicitly accords priority to the "liberty principle" over those other maxims that point toward greater equality.

is worth stressing that equality as I have described it does not stand alone, but is closely related to the idea of liberty. The relation is complex, and I cannot say very much about it here. It is a feature of the argument I have made, however, that the right reason for distributing love, belief, and, most important for my immediate purposes, political power is the freely given consent of lovers, believers, and citizens. In these sorts of cases, of course, we all have standards to urge upon our fellows: we say that so and so should not be believed unless he offers evidence or that so and so should not be elected to political office unless he commits himself to civil rights. But clearly credence and power are not and ought not to be distributed according to my standards or yours. What is necessary is that everyone else be able to say yes or no. Without liberty, then, there could be no rightful distribution at all. On the other hand, we are not free, not politically free at least, if *his* yes, because of his birth or place or fortune, counts seventeen times more heavily than *my* no. Here the case is exactly as socialists have always claimed it to be: liberty and equality are the two chief virtues of social institutions, and they stand best when they stand together.

(1973)

16

Democratic Schools

Education is not a good that we can simply distribute to needy or deserving children, like money or food, housing or health care. It must be received as well as given, and a growing amount of evidence suggests that the quality of the reception has little to do with the quantity of the giving. If we could concern ourselves solely with the allocation of educational resources, it would not be difficult to work out distributive principles: schooling ought to be provided for each child up to the limit of his capacities, or short of that, for as long as he wants it. And it should be equally provided, at least in the limited sense that, within the public system, roughly the same amount of money should be available for each child (I leave aside here the additional help that must obviously be provided for handicapped or disabled children). No doubt, a great deal remains to be done to fulfill these principles; they would require significant changes in the financing of public education, for example, so as to eliminate the dependency of local school districts on local economies.

But the major problems that beset primary and secondary education in the United States today are not problems of distribution and cannot be solved by redistribution. I want to

257

suggest that we think of them as problems of association. In a pluralist society, the preparation of children for political and cultural citizenship requires the creation of communities of learning, communities that have no *a priori* existence and must be constituted by political decision. The relevant decisions are not only those currently required by disputes over racial integration. There are general questions about integration and segregation that have to be answered. Which children should go to school with which others? On what principles should we associate children in schools and classrooms?

Just as these are not questions of race alone, so they are not questions only of social class. They would arise even in a classless society, unless we assume (as many socialists, particularly in the Marxist tradition, did assume) that classlessness would also involve the submergence or overcoming of religious, ethnic, and ideological differences. If socialist citizens had only that one, socialist identity and no other, the association of their children might raise no problems. But I expect that if they were good socialists, and typical ones, they would find things to disagree about. They would certainly disagree about educational philosophy, and so the question would inevitably arise: should children whose parents are philosophically divided attend different schools?

The search for an associative principle has reference only to children. Except in armies and juries, adults are not coercively brought together by the state. Their range of acquaintance and activity may be limited in all sorts of ways, but with reference to the state they associate freely, following whatever principles they choose. Freedom is the ideal here (and whatever degree of "friendship" is necessary to democratic politics), but this is the freedom of particular men and women, with formed and stable identities they have not freely chosen. These identities are in part the product of their education—probably less of the formal curriculum than of the educational environment and the associations it requires—and in part the product of the same nongovernmental factors that presently limit their social range (and previously limited the social range of their parents).

Conceivably, we could maximize and equalize adult freedom by removing children at an early age from every sort of social influence and randomly associating them in identical schools. But the products of such an education would not be particular men and women in our present sense; nor would the society they formed be a pluralist society. Random association would represent a radical triumph of state and school over society and family, and though something like it has occasionally been advocated by leftist groups, it could probably be enforced only in a totalitarian regime.

Short of random association, what is the right associative principle? I can think of five possibilities, each of which I want to examine in some detail. They point toward different though overlapping conceptions of what the educational experience should be like, and the fact that there are five suggests our present confusion on that issue.

In weighing the five possibilities, I shall adopt the following guideline: *the principle on which children are coercively associated should anticipate the pattern that would prevail among adults in a world of freedom and equality.* In an instrumental sense, the anticipation may not do any good; it may not bring adult freedom and equality any closer—for that requires, whatever we do with children, the political activity of adults. But the school years are long, a substantial part of a human life, and they have a value in themselves, without regard to their long-term results. We want what is best for our children right now. Moreover, because of this parental concern, educational debates have significant short-term results: they are likely to generate adult political activity of a very intense kind. Christopher Jencks has suggested that the most important effects of school desegregation are on adults, not children.[1] It's not that associating children from different racial groups brings their parents together; it doesn't. But it does force parents to think about associative principles, and these quickly open up into larger political issues.

[1] *Inequality: A Reassessment of the Effect of Family and Schooling in America* (New York, 1972), p. 156.

First principle: neighborhood. There are such things as natural or spontaneous neighborhoods, the free creations of people who know one another and choose to live together. But most neighborhoods are political artifacts. Settled in chronological layers by different groups, their boundaries are the results of zoning laws, highway placement, economic development, subway and bus routes, and the location of public schools. A neighborhood school, then, doesn't necessarily serve a homogeneous community, though insofar as different groups of people come to view a particular school as theirs, it may serve to heighten feelings of homogeneity. This was one of the original purposes of the public school: it was to be a little melting pot, and neighborhood was to be the first of its products, on the way, as it were, to citizenship. It was assumed that school districts geographically drawn would in fact be socially mixed, that the children who came together in the classroom would come from very different class and ethnic backgrounds. This was never consistently true, across any particular city or town; I am not sure whether it is more or less true now than it used to be. With regard to racial mixing, however, the evidence is clear. Neighborhood schools serve to keep black and white children apart. Indeed, keeping them apart was one of the chief reasons for the political decisions, above all the gerrymandering of school districts, that constituted the neighborhoods. For this reason, the associative principle of neighborhood is today under severe attack. Nevertheless, it remains for many people an attractive idea, quite aside from racial issues, and it is worth explaining why.

If one tries to imagine the basic units of a pluralist society, the geographic district is hard to avoid. People are most likely to be active and effective where they live. It is in their neighborhoods that they are knowledgeable, involved, directly concerned—and the claim that children should be educated in that world where their parents are most at home is very strong. It is strong, above all, because parents make the claim so strongly, because the schools are, as I've already suggested, such an important focus of local politics. But we should pay attention to it for other reasons. Children probably do better in schools their parents support; parents probably do better, as parents, when their children are going to schools whose character they understand and accept. I don't want to deny the role of the school as an alternative environment, a liberation from home life, a potential source of conflict in the socialization of children. But it is another of the advantages of the local school district that it is usually large and diverse enough to allow for that sort of tension. Since decisions about local public schools are political in character, they necessarily involve bargaining and compromise. Few parents will be entirely satisfied (unless they are entirely passive), and children are almost certain to find a world at school different from the one they know at home. Different, but not too different—so this same school can serve as a center for neighborhood activities: political meetings, athletic contests, concerts, dances, extracurricular classes in everything from car repair to yoga.

I've made the picture too "pretty." It is necessary to add that this first associative principle works better for some Americans than for others; it works best for those who live in cohesive and relatively prosperous neighborhoods. But there can be no doubt that most Americans would live in such neighborhoods if they could—and send their children to the local schools, involve themselves in school board elections, and so on. Before turning, then, to the problems raised by our actual residential patterns, I want to take up two alternative principles that arise among people for whom local schools are already effective, and that would arise even if all American neighborhoods were cohesive and prosperous (integrated, classless, or whatever).

Second principle: parental interest or ideology. We don't, in fact, bring all children together in public schools; we permit their parents to send them to private schools, supported (largely) with private money. Parents with sizable incomes or parents belonging to organizations, like the Catholic Church, with substantial resources, can associate their children in accordance with religious conviction, social standing, or educational philosophy. Among middle-class Americans there has recently been some proliferation of "free schools" (which are often very expensive) organized on the last of these criteria. The products of the affluent sixties, they may well collapse in the seventies, but they clearly reflect a growing sense that parental interest ought to prevail over democratic decision making with its inevitable compromises. Arguments for neighborhood control would seem to apply *a fortiori* to private schools insofar as these are especially exposed to consumer pressures. Why shouldn't parents get *exactly* what they want for their children? At present, only some parents can do that, but the ability to do it could be widely extended if public support were forthcoming. This is the thrust of the "voucher plan," a proposal that tax money available for education be turned over to parents, in the form of educational vouchers, to be spent on the open market in whatever schools they choose. These would not be neighborhood schools, except insofar as parents with similar educational commitments happened to be living together. In any case, they would be oriented more precisely to parental wishes than neighborhood public schools are ever likely to be. There would also be more of them, presumably, and they would cater to a wider range of interests and ideologies.

The voucher plan is a pluralist proposal, with the basic unit shifted from the geographic district to the self-forming community of like-minded parents. Or, more accurately, it points toward a society where there would be no basic unit at all, but a variety of units for different sorts of activities. It might be argued that this is the pattern that would emerge among free and equal adults: they would be highly mobile, rootless, moving easily from one ideological community to another; they would make different decisions in different groups and so avoid the endless log-rolling of geographically based politics.

I am inclined to be dubious about this picture, but in any case it does not suggest the experience children would have in schools freely chosen by their parents. For most children, parental choice almost certainly means less diversity and less tension. Their schools would be more like their homes than they are now, and the children they were associated with would be more like themselves. Perhaps these arrangements predict their own future choices, but they hardly predict the full range (the desirable range) of their future acquaintance. Ideologically based schools might cut across ethnic and racial lines in a way that neighborhood schools often do not. But even that is uncertain, for ethnicity and race would surely be two of the principles around which free schools were organized. These are entirely acceptable principles, so long as they aren't the only ones, in a pluralist society. Once again it has to be stressed, however, that for particular children they would be the only ones.

The most disturbing feature of the voucher plan is its reliance on the open market. Once vouchers are in circulation, we must expect the appearance of large numbers of entrepreneurs, with educational goods to sell. And since many parents are passive and confused with regard to their children's education, they (and their children) are ripe for an old-fashioned kind of capitalist victimization. In public schools, such parents are often protected by the political activists in their midst, who serve the interests of the community as a whole by keeping some pressure on teachers and administrators. In a free school system, interested parents would more likely vote with their feet, withdrawing their children from schools whose value they had come to doubt, leaving behind uninterested parents and helpless children. There would be little pressure on educational entrepreneurs, so long as they could replace departing customers—except insofar as governmental inspectors made their appearance, enforcing some general code.

Despite all this, there is much to be said for the argument

that parents not be locked into their neighborhood public schools, or rather, that their right to exit not be conditional on their financial resources. After all, voting with their feet is often the only way they can vote that is immediately effective for their own children. Political action on educational issues tends to be especially altruistic; its beneficial results most often benefit other people's children. More immediate parental concerns might be accommodated by permitting what is called "open enrollment" within the public system. Given such an arrangement, most parents would probably prefer neighborhood schools, and prefer also to fight for better schools in their own neighborhoods, but there would be a way out for those particularly dissatisfied. Or, at least, there would be a way out if different neighborhoods had different sorts of schools. I will return briefly to this possibility later on, in talking about racial integration; I want to consider now another associative principle that parents often seek to implement either within or in spite of neighborhood schools that work.

Third principle: talent. The idea that careers should be open to talent is central to liberal society, and it has often been argued that school careers should be similarly open. Then those children who can move along quickly will be enabled to do so, while slower students will find their work adjusted to the pace of their learning. Both groups will be happier, so the argument goes, and within each group children will find their authentic friends and future allies. In later life, they will continue to associate (and would do so even in a world of freedom and equality) with people of roughly similar intelligence. Parents who think their children especially bright tend to favor this sort of segregation, partly so that the children make the "right" kind of friends, partly so that they are not bored in school, partly in the belief that intelligence reinforced is even more intelligent. Just for this reason, however, there is often a counter-demand—that bright children be distributed throughout the school or the school system so as to stimulate and reinforce the others. This looks like using the bright students as a resource for the less bright, a use we would not permit among free and equal adults. But whether it is "using" or not depends upon what one takes as the natural starting point for coercive association. If the starting point is everyday residence and play, then it is the segregation of the bright students that can plausibly be criticized: it now seems to be a willful impoverishment of the educational experience of the others.

It's not really true that the adult world is voluntarily segregated by intelligence. All sorts of work relationships, up and down the status hierarchy, require mixing, and what is more important, democratic politics requires it. One could not conceivably organize a democratic society without bringing people of every degree and kind of talent and lack of talent together— not only in cities and towns but also in parties and movements (not to speak of armies and bureaucracies).

Schools should anticipate that kind of integration, and only when they do so, it seems to me, are more limited uses of segregation permissible. There are educational reasons for separating out children who are having special difficulties in math, for example, and also children who are especially good at math. But there are neither educational nor social reasons for making such distinctions across the board, creating a two-class system within the schools or creating radically different sorts of schools. When this is done, and especially when it is done early in the educational process, it is not the voluntary associations of adults that are being anticipated, but the class system in roughly its present form. For then children are brought together chiefly on the basis of their preschool training and their home environments, in ways that drastically limit their future development. Moreover, in contemporary America, they are brought together so as to establish a hierarchy not only of social classes but also of racial groups. Inequality is doubled, and the doubling, as we have reason to know, is especially pernicious for democratic politics.

We can try to minimize the coercive effects of the class

structure and of racism in two ways: first, by avoiding every kind of premature labeling of students and keeping channels of opportunity open as long as possible; second, by building a more fraternal pattern of association in our schools (but not only there). The claim that neither of these goals can be achieved within neighborhood schools, even if they are un-tracked, generates the fourth associative principle.

Fourth principle: equal treatment. The argument here is that neighborhood schools are radically unequal, not because or not primarily because of the amounts of money spent or the quality of the teaching or the content of the curriculum, but because of the social character and expectations of the children associated within them. In ghetto and slum schools, children are prepared for ghetto and slum life. The world they see, the contacts they make, the debased value of the certification they receive: these are the crucial features of their education. In effect, they are labeled by their social location and taught to label themselves. Their schooling is from its first moments an attack on their self-esteem, and their performance in school reflects the success of the attack. The only way to alter this situation, it is said, is to shift the social location of the students, that is, to separate their schools from their neighborhoods. This can be done both by moving ghetto and slum children out of their local schools and by moving other children in. Either way, it is the associational pattern that is being changed.

It's not easy to say exactly what new patterns the principle of equal treatment requires. We are likely to be pressed toward a system where the social composition of each school would be exactly the same. Different sorts of children would be mixed in the same ratio in every school within a given area, the ratio varying from area to area with the overall character of the population. But how does one identify the appropriate areas? And how does one sort out the children—by race alone, or by religion too, or by ethnic group, or social class? Logic would seem to require the largest areas within which transportation is possible and the most detailed sorting out. But the federal judges who are currently deciding such questions seem likely to stop short of both, focusing their attention on cities and (perhaps) their immediate suburbs and on racial integration alone. "In Boston," Judge Garrity declared in a decision requiring extensive intra-city busing, "the public school population is approximately two-thirds white and one-third black; ideally every school in the system would have the same proportions." No doubt there are good reasons for stopping at that point, given the special role of race in American society, but it is worth emphasizing that equal treatment, taken literally, would require a much more complex proportionality.

Because of the current stress on race, arguments against proportional association often take racist forms. I don't think they need do that; it is only necessary to suggest that proportionality would not characterize the relations of free and equal adults. Among certain black activists, for example, the argument goes this way: even in a society free of every taint of racism, most (or at least many) black Americans would choose to live together, shaping their own neighborhoods and controlling local institutions. The only way to anticipate that social pattern is to establish neighborhood control now. If increasing numbers of black adults (parents as well as professional teachers and administrators) take an active part in the schooling of black children, if black communities become centers of educational activity, the ghetto will cease to be a place of discouragement and defeat. What equality requires, in this view, is that the association of black children with other black children carry with it the same mutual reinforcement as the association of white children with other white children. To opt for proportionality now is to acknowledge the impossibility of such reinforcement, to admit the failure of black community—and to do so before any serious effort has been made to succeed.

This is a powerful argument. It draws force from the principle of neighborhood, which, as I have been trying to indicate, must be the preferred principle in a democratic and pluralist society. But it faces in America today a major difficulty. The residential segregation of black Americans is very different from that of Irish Catholics, WASPs, Jews, and so on. It is, to put it simply, a great deal less voluntary and a great deal more thoroughgoing. It does not anticipate pluralism so much as separatism; it is not the pattern that we would expect to find in a racially open society. Surely we ought to avoid associating children so as to consolidate that pattern. But how far can we deviate from it while still respecting the communities that blacks would form in an open society? Equally important, how far can we deviate from it while still respecting the communities that other people have already formed? I don't know exactly how to draw the line, but I am inclined to think that proportional association draws it badly.

If equal treatment means identical schools, then the principle is incompatible with the existence and development of a pluralist society. For as long as adults associate freely in groups, shaping diverse communities and cultures, the education of children will be, and should be, group-dependent. But adults don't associate freely enough: above all, in the United States today, contacts between the races are severely constrained. Hence the schools must try to bring children together in ways that challenge the existing barriers and maximize the possibilities for cooperation and mutual involvement among members of different groups. For this purpose, it is not necessary that all schools be identical in social composition but only that different sorts of students encounter one another within them.

This probably requires what is currently called (by those who oppose it) "forced busing." The phrase is unfair, since all school assignments are compulsory in character, but it contains a grain of truth, and it suggests why busing, however necessary, is not an ideal procedure. Though children will undoubtedly make the best of it, if left alone, it represents a more overt kind of coercion, a more direct disruption of everyday living patterns than is desirable. Under the best of circumstances, there is sure to be tension and conflict when groups of children who live entirely apart are suddenly thrown together. More centralized patterns of political control are likely to be necessary once the principle of localism has been set aside. On the other hand, it is clear that that principle has been set aside before—by school committees and city councils that imposed racial segregation even when geography called for, or at least allowed for, very different associational arrangements. Given the actual histories of many school districts, further assaults on the norms of neighborhood seem unavoidable, though I would seek to minimize them wherever possible. Other arrangements should also be explored: parental choice within the public system (open enrollment) may operate against racial segregation and should certainly be encouraged if it does so; the creation of "magnet schools" catering to particular student interests may have a similar effect. And more desirable patterns of student contact could probably be achieved by giving up the requirement that children (particularly older children) spend thirty to thirty-five hours a week in the same school building. It might be preferable to bus children for parts of days, associating them in different ways for different activities, while maintaining a neighborhood base. But it's not my purpose here to offer practical suggestions, only to recognize the force, and also the limits, of equal treatment as an associational principle.

Fifth principle: nationalism. Given today's residential patterns, the ideology of the melting pot probably requires the same kind of racial (and class) mixing as does the principle of equal treatment. The requirement is not, however, equally strong. The idea of a nationalist education drew whatever strength it once had from the perceived difficulties of immigrant absorption. I don't believe that contemporary forms of ethnic assertiveness pose the same or even remotely similar difficulties. Failures of democratic socialization today have more to do with the weakness or inadequacy of local and particularistic institutions than with the absence of a common language or a national consciousness. It's hard to imagine any

group of children brought together under any possible circumstances in America today who would not be recognizably
American, to themselves as well as to other people. They don't
need any further "melting," but they do need, and schools
should provide, an environment within which they can take on
some independent shape.

That shape will of course be mediated by their homes and
neighborhoods as well as by their schools. But I see no reason
for state or federal governments to take a direct role in
establishing the sovereignty of the school (beyond the role
they already take in setting minimal standards). No doubt, they
could make American nationality a more uniform product if
they did intervene, particularly if they did so by separating
children from their homes and neighborhoods. That might
also be a way of reducing racial and ethnic prejudice—if only
because children who came increasingly to resemble one another would have to find new ways of disliking one another.
But it is not an attractive prospect, for what is valuable and
distinctive about America is precisely its pluralist character:
the absence of nationality in the usual sense of that term and
its replacement by citizenship. Since the association of free
and equal citizens in a pluralist society is determined in large
part by their group identities, there is no reason to refuse to
allow similar determinants to work among children. They won't
be the only determinants at work, for the fact that the schools
are preparing children to participate in a single economic system and a national politics assures a certain uniformity in
educational outcomes. The omnipresence of the mass media
and of the cultural products they purvey heightens the as-
surance. Pluralist tendencies play against these forces, I think,
in a salutary way; in any case, the play is unavoidable unless
the government were to commit itself to a *kulturkampf* of
extraordinary proportions.

A rather simple and, I am afraid, unoriginal proposition
follows from the survey of associational principles I have just

completed. The most important battle for democratic education was won when public schools were established and given over to local control. Educational politics has, since then, been the most decentralized form of American politics: I mean to endorse that decentralization. The phrase "local control," of course, begs many questions. In fact, the localities are radically different in character and size, and I have said nothing about the desirable shape of school districts or the proper authority of school committees. These are matters (sometimes) worth fighting over—along with finances, curriculum, and so on; but the structure of the inevitable battles is set and, I think, rightly set by the present patchwork of public systems. The existence of that structure is already a victory for democratic schools, and we should not be quick to relinquish that victory even when, because of failures elsewhere in society, we have to compromise its character.

Today we are involved in a major compromise that is likely to require larger school districts, centralized decision making, increased use of busing, and so on. These are expedients made necessary by prevailing residential patterns (and by past school districting decisions), but they are not desirable in themselves; they should be temporary expedients; in a just society no one would have any reason to suggest them. When men and women live as equals in communities they have chosen and over which they exercise some significant degree of control, they will, by and large, want to educate their children where they live—and that preference should be respected as a matter of course.

They will disagree about what should go on in the local schools, and these disagreements will be an important part of local politics (and local politics will be important because of them). Across a city or state, very different decisions will be made in different school districts. Some of these districts will be relatively homogeneous in character; some will be radically heterogeneous. Sometimes the boundary tensions endemic to a pluralist society will have to be faced in a particular community or a particular school, but we need not insist that this be the case everywhere and all the time. Most often, within particular districts parents with diverse commitments will establish a *modus vivendi*. Within particular schools children will have to cope with the common but not, or not often, with the extreme differences that pluralism makes possible.

This is, I think, the natural form of democratic politics and of democratic education. It invites participation within a familiar world; its institutions are built to a human scale. It opens the way to every form of diversity while still permitting people who choose to do so to live and work together—and to bring their children together. It promises local excitement, well short of civil war.

(1976)

17

Town Meetings
and Workers' Control:
A Story for Socialists

Introduction to the Story

There are thirteen arguments for socialism; they have to do
with distributive justice, equality, the need for planning, self-
respect, fraternity, and so on. But the one that seems to me the
easiest and best is a political argument, an extension of the
defense of democracy. It has been put forward often over the
last one hundred years, but it has never, in this country at
least, commanded general acceptance. I suppose no doctrine
commands general acceptance that is, as Hobbes wrote, "contrary to any man's right of dominion or to the interest of men
that have dominion."[1] And yet there is some sense in which

[1] *Leviathan*, Part I, chap. 11.

we are all democrats, so I shall start from there, assuming that we have good reasons, and see how far I can go.

The central commitment of socialist politics has often been put in a phrase that must be intuitively appealing to democrats: *the abolition of the power of man over man.* Neither democrats nor socialists begin with an assertion of popular sovereignty. Since they everywhere encounter established sovereigns, authorities, hierarchies, conventional claims to rule, they begin with denials and rejections. They are abolitionists. They aim at abolishing two kinds of authority relations, those in which men and women are directly, and those in which they are indirectly, subject to the arbitrary will of another. I will consider these two separately. Direct subjection suggests the immediacy of bondage; it describes the slave, the serf, the servant, all those who bow before some powerful person, defer to him, obey his every command. Direct subjection is pervasive in the old regime, and it is not missing in the new. One recognizes it in those forms of speech, those bodily postures and motions that connote weakness, inferiority, humility, a certain zeal for service, which we are disinclined to accept as spontaneous or voluntary.

Indirect subjection is not so easy to recognize, for it has to do not with relationships but with systems of relationships, and the systems are invisible. What is at issue here is the right of a single person, acting on his own, for reasons of his own, to make decisions seriously affecting the welfare of his fellowmen, without the agreement of those whom his decisions affect. Now we all make decisions all the time that seriously affect others: when I decide to accept a job, for example, my decision has an immediate impact on the life of the next candidate, who would have received an offer had I declined. But it would be odd to think of him as my subject, and I am certainly not required to seek his agreement before making up my own mind. In every society, however, there are positions of recognized power, offices within some organizational structure, from which decisions of a different sort are made. The best way to characterize these is to say that they are authoritative decisions, for that invites us to inquire as to the source of the authority. And what is crucial in systems of indirect subjection is that the subjects are not the source. Men and women ruled by an absolute monarch are directly subject in that they must obey his commands and indirectly subject in that they must live with the consequences of his political, economic, and military decisions. Clearly, democratic argument must challenge and does challenge both these forms of subjection.

The principle of the second challenge is nicely expressed in an old maxim, a rule of law, I believe, in medieval times, honored regularly in the breach: *what touches all should be decided by all*. It's not every decision affecting others that must be democratically made, but only those affecting everyone. That does not mean everyone in the world; I think we should take it to mean everyone associated in some common enterprise the existence or success of which requires that decisions be made. Thus the medieval maxim was invoked in disputes about authority relations in guilds, churches, towns, and states. No doubt, there are problems with this (and every other) restriction on the reach of the maxim, for decisions made within some particular association are certain to affect not only members, but nonmembers as well—and may affect them seriously indeed, as the example of state decisions and international politics suggests. But since this is a general difficulty for democratic theory, and not a special difficulty for socialists, I am going to put it aside (I'll come back to it briefly later on). My own argument requires only this much: that for any authoritative decision seriously affecting others, there is some subset of those affected who ought to authorize the decision or to make it or approve it themselves. The subset includes, but is not necessarily limited to, those people whose association makes the decision possible or necessary—as the association of Frenchmen, for example, makes possible the French king's decision to go to war or requires him to oppose an armed invasion.

The argument also assumes that the king cannot decide to

break up the association (or to cooperate with the invaders) rather than allow its members to join him in decision making. It is not the case that the authorities can seek support for their decisions just anywhere. Thus Bertolt Brecht's satiric poem, written in Berlin, 1953:[2]

> After the rising of the 17th of June
> The Secretary of the Writers' Union
> Had leaflets distributed in the Stalinallee
> In which you could read that the People
> Had lost the Government's confidence
> Which it could only regain
> By redoubled efforts. Would it in that case
> Not be simpler if the Government
> Dissolved the People
> And elected another?

The satire depends on the commonly accepted principle that the dissolution of the government is a popular right. But it should also be said that the dissolution of the *people* is a popular right. The cohesion that makes it possible to speak of "touching all" and being "decided by all" can only be dissolved by all. And this is true however the association was made and whatever the role of the king or the authorities in making it. The argument is not dependent upon an original contract; it works whatever stories about the origin and foundation of the common enterprise are accepted by the participants, including stories about great men, godlike acts, vanguard revolutions, and so on.

Now, it is the socialist claim that neither direct nor indirect subjection, neither the arbitrary power of persons nor that of positions has been abolished through the establishment of a democratic state. For the state is not our only common enterprise; nor do the laws of the state constitute the only disci-

[2] Quoted in Martin Esslin, *Brecht: The Man and His Work* (Garden City, N.Y., 1961), p. 183.

plinary system to which we are subject. The capitalist economy proliferates what are plausibly called private governments. Within capitalist organizations a process of decision making can be marked out, dominated by officials, which has the crucial characteristics of a political regime. The process has outcomes that seriously affect thousands and hundreds of thousands of people, including men and women whose co-operative activity underlies the organization and who are in some sense its members. These outcomes take the form of decisions and rules that can be opposed or ignored by the members only at the risk of penalties. Here are participants who are subjects, officials who act with authority. What is the source of that authority? It clearly does not derive from the participants, else it would not be called private. It derives instead, or it is said to derive, from the ownership of the organization by particular persons. Their claim to govern, to make decisions affecting others, rests on the legal and, what is more important, the moral implications of private property. Socialists argue that this is not a tenable claim, and it is at this point that many democrats part company with them, insisting that economic enterprises are unlike political associations precisely because the former are subject to ownership and the latter are not.

It should be said, however, that in their time democrats also challenged the implications of ownership—as these were understood within the feudal economy. Feudalism, like capitalism, rested on a certain view of property rights, specifically on the view that the ownership of land entitled the owner to exercise direct disciplinary (judicial and police) powers over the men and women who lived on the land and also to make decisions (to go to war with some neighboring landowner, for example) seriously affecting their lives. In the course of many years of political conflict and revolutionary activity, the formal structure of feudal rights was abolished and the disciplinary powers of the feudal lords were socialized. Taxation, law enforcement, conscription: all these ceased to be property rights. In Marxist terms, the state was emancipated from civil society, that is, from the property system. The implications of ownership were redefined so as to exclude certain sorts of decision making that, it was thought, could only be authorized by the political community as a whole. This redefinition establishes the central division along which social life is organized today. On the one side are activities called political; on the other, activities called economic. On both sides, men and women make authoritative decisions affecting others, but the maxim *what touches all should be decided by all* applies only in the realm of politics. Hence socialism has commonly been described as the extension of democratic decision making from the political to the economic realm.

But this description may be misleading, for the two realms do not seem to me at all distinct. They are subject, of course, to conventional definition, but there is no need for us to accept the conventions of 1789. Indeed, the political argument for socialism is strongest insofar as it suggests the radical similarity of decision making in the two realms. What justifies the contemporary version of property rights, we are commonly told, is the entrepreneurial zeal, the risk taking, the inventiveness, the capital investment, through which economic enterprises are founded, sustained, and expanded. Whereas feudal property was founded on armed force and sustained and expanded through the power of the sword (even though it was also traded and inherited), capitalist property rests upon forms of activity that are intrinsically noncoercive. The factory is distinguished from the manor, the disciplinary system of the first is upheld and that of the second condemned, because men and women come willingly to work in the factory, drawn by the wages, working conditions, prospects for the future that the owner offers, which are made possible by his energy and enterprise, while the workers on the manor are serfs, prisoners of their noble lords. All this may well be true; in any case, I will not question it now; it helps us understand why feudalism was not an ideal political or economic arrangement. But it does not draw the line between democratic politics and the capitalist economy, nor does it justify the present authority of owners. For political communities are also created by entrepreneurial energy and enterprise, and it's not implausible to say of cities and towns, if not always of states, that they recruit and hold their citizens by offering them an attractive place to live. Yet ownership is not an acceptable source of governmental authority in cities and towns. If we consider deeply why this is so, we will have to conclude, I think, that it should not be acceptable in companies or factories either. What is necessary is to imagine a man who claims to own a town, to tell a story—it must be a success story—about the life of a political entrepreneur.

The Story

Long ago, when the frontier was still somewhere east of the Great Plains, a young man named J.J. set out to make his fortune. He was bold, adventurous, energetic, and very smart, and he left Boston and New York, even Pittsburgh and Cincinnati behind him. After hardships and excitements not worth mentioning, he staked out a claim to a large and rich piece of land at the bend of a river, one of the smaller western tributaries of the Mississippi. But J.J. was not a farmer. He hated domestic animals, and while he could plow as straight a furrow as anyone in the West, the accomplishment gave him no joy. Soon, he acquired land on the other side of the river and built a ferry, which he ran himself. It was a well-built and well-run ferry. J.J. was a gregarious man, and he entertained his passengers as he took them across; he was warm and funny and full of stories. The enterprise was a success, and after a year or so a few men settled nearby, a storekeeper, a blacksmith, even a preacher, renting the land from J.J. He provided a small lot for a small church and happily

watched the settlement grow. When word came of a threatened Indian attack, he organized its defense, bringing in (and
paying for) vital supplies from the East. There was an attack
of sorts, though only by a small raiding party, and J.J. was
in the forefront of the defenders—twelve or fifteen people
now, mostly heads of families, who recognized and accepted
him as their leader. He would have been the newest of new
men in Boston or New York, but in this little settlement, he
was the oldest, the richest, the most well-established of the
inhabitants.

About this time, J.J. went East to borrow some money.
He now had visions of a city, for the river bend was a good
location, the ferry was busy, there were new farms on both
sides of the river, and the farmers needed supplies, schools,
sermons, and company. He got the money from a young banker
who was bold, adventurous, energetic, and very smart. Back
home, he bought more land, laid out a square, set aside lots
for a school and a town hall. Though no one had yet died in
the settlement, he provided a cemetery too, sure that people
would die and not unhappy about that. Dead bodies lend dignity to the place where they lie; a town with a cemetery has
staying power. He incorporated the town in accordance with
territorial law and gave it the name he had always had in
mind: J-town. The law did not specify any particular form of
government, but J.J. again had something in mind. When the
town hall was built (at his own expense), he moved in. The
settlers were not surprised; nor was there any opposition. J.J.
was still a gregarious man; he knew them all, talked to them
all, always consulted with them about matters of common
interest. As at the time of the Indian raid, his leadership was
recognized and accepted. Anyway, they all paid him rent, and
it did not seem strange to pay him taxes too—a per-capita
levy for the salary of the schoolteacher, the maintenance of
public buildings, and other minor expenses. J.J. himself took
no salary. His ferry was doing well, and he had begun to
transport goods up and down the river. His enterprises ex-

panded as the town grew and, to a considerable extent, the town grew because his enterprises expanded.

Years went by. J.J. prospered, paid off the loan, delighting himself and the Eastern banker. J-town prospered; new settlers arrived every year; there was money in the treasury. The town lines were redrawn. Now there were tax-payers who did not pay rent to J.J., though he still owned most of the town and continued to serve as mayor. When it became necessary to appoint other town officers, he talked to his friends and neighbors and always picked the right person. He had a knack for making decisions, not only decisions that paid off, but ones that pleased people. And if anyone wasn't pleased, he could always move on; J.J. would pat him on the back and talk about the wide Western spaces. He treated criminals the same way. He didn't like locking people up, and he thought it enough of a punishment to have to leave J-town.

J.J. was a natural leader, a man of substance, a man of power. He proved himself again when the flood came, risking his boats on the rapidly rising river in order to evacuate the settlers and their movable possessions, contributing freely to the relief fund. Later on, he went East for a second time to raise money, signing the papers in his own name and bringing home the capital that made it possible to rebuild J-town. The townspeople could hardly imagine another mayor. Nor could J.J.

Life was placid in J-town; time passed quickly. Revivalist preachers came and went; a labor organizer arrived one day and departed the next; the Republican party sent someone to talk to J.J., and the visitor did not find it necessary to talk to anyone else. The dissidence of dissent was absent. Or so it was until J.J., aging now, appointed his son chief of police. Perhaps there had been murmurings before that, but they were scarcely audible. Many of the new settlers did not remember J.J.'s earlier days, did not really believe the stories of his heroic exploits, or—what was worse—did not think the stories mattered. But the town was well run, and they were content not to worry about it; they didn't think of politics as something that was missing from their lives; to all appearances they didn't think about it at all.

Still, the appointment of his son was a political mistake—J.J.'s first. His son wasn't particularly bold, adventurous, energetic, or smart, and everyone in town knew it. A few of the newer inhabitants called a meeting at the Odd Fellow's Lodge (the town hall had no assembly room). Only a small number of people came, but they shouted a lot, worked one another up, formed a citizen's committee, and called another meeting. This time more people came; the speeches were exciting, and the participants went home feeling differently about J-town than they had ever felt before. There was a third meeting, and on the following afternoon a delegation of citizens called on J.J.

No one took notes that afternoon, but what was said was repeated all over town. The citizens told the mayor that he could appoint his son to any position he liked in J's River Haulage, Inc., but the town, they said, was not one of his businesses. Town government was the *public* business, and henceforth the public would have to be brought into it in some regularized manner. What touches all should be decided by all, they said. They demanded that elections be held, and they intimated that they had a candidate for mayor in mind whom they preferred to J.J.

J.J.'s reply was a passionate defense of entrepreneurial rights. What do you mean? he said. *This is my town.* I found this place; I built my ferry here, and other people came because of the ferry. I risked my life against the Indians. I risked my capital to buy this land and raise on it the buildings people needed. Your children are studying, right now, in schools I made possible; your dead are buried on land I gave. When the flood came, I did it all again, giving money, raising money, organizing reconstruction. This town wouldn't exist without me and, what's more, I still own most of it. All of you came here with your eyes open; you knew how this town was run; you knew who made the decisions around here. Don't talk to me about elections. . . .

J.J. had never lacked for words. And everything he said was true. He did not have to make things up, for he was indeed a great man and had done great things. If it were possible to own a town, he certainly deserved to own this one. The citizens insisted, however, that it wasn't possible. J.J. was the founder, not the owner of J-town, they said, and they would happily put up a statue of him in Central Park. (On land I gave you, muttered J.J.) He was entitled to honor and glory, but not to obedience. He was entitled to rent, but they would tax themselves. They did not mean to sound ungrateful, they said; they weren't ungrateful, but political foundation and public service did not give a man the right to tyrannize over others.

How can I be a tyrant, shouted J.J., when it's my own town?

The next weeks were exciting, but the events are not worth mentioning here. The citizens' insurrection was successful. J.J. fell. His son never became chief of police. J.J. withdrew from public life, did not vote in the first town elections, never attended town meetings, turned away from his friends and neighbors. When he died, not long after, the town council commissioned a statue, as the insurgent citizens had promised. The new mayor made a fine speech when the statue was unveiled. J.J., he said, was a heroic figure, a man of the pioneering past, a founding father. The town owed him a profound debt, a spiritual debt, of which the statue could only be a lasting reminder. It was true, the mayor said, that J.J. had occasionally confused business and politics, but, after all, that was a confusion not uncommon in American life.

J.J.'s son had occasion to reflect upon that last remark when the town council, only a few years later, tried to take over the river haulage company. They would have paid him handsomely, had the state supreme court allowed the take-over, but money, they insisted, was all he was entitled to in return for his father's investment. Decisions about whether or not to expand the ferry service, buy new boats, and so on—those could not be made by one man, they said, considering only his private profit. The whole town was a transportation center; all its citizens depended on the haulage company; most of them worked for it. And what touches all, they said, should be decided by all. . . .

Interpretation of the Story

I had originally planned to write about a company town, drawing upon actual historical accounts. The story would then have described how the owners of an economic enterprise created a political community, so to speak, on the side, as a place for their employees to live. Upon reflection, it seemed better to imagine a case where the major entrepreneurial activity was focused on the town itself, so that J.J. would stand in the great tradition of the political founder. He represents the liberal form of that tradition, which is to say, he has no deep convictions about the shape of the town or the moral character of its citizens. He merely wants them to live peaceably and to prosper. But he is clearly comparable, despite that, to the figures celebrated by Machiavelli and Rousseau—Lycurgus, Solon, Romulus—who made or remade the political community. Now, in that tradition, founding or reforming the state generates no right of ownership and none of the subsidiary rights that ownership brings with it in feudal manors or in capitalist factories or companies. Most important, it does not give rise to any sort of disciplinary authority over those who join the new community. That authority belongs to the members, even if they were passive or entirely absent during the period of foundation.

Why are economic associations any different?

Not because of the entrepreneurial vision, energy, inventiveness, and so on, that go into the making of the company:

the making of J-town required exactly the same qualities. There are certainly objects in the world that a man can acquire through enterprise and invention. Making them or mixing his labor with them, as John Locke argued, produces at least a presumption of ownership.[3] But it's not the case that *anything* can be acquired that way, and if towns cannot be, there is no reason to think that companies can. These two are much nearer to one another than either is to the land I cultivate, the wood I cut, the chair I build, the book I write—the examples that Locke had in mind. What brings them close together is that both of them involve other people, shared interests, cooperative activity.

Not because of the investment of capital: J.J. invested his own money in J-town without becoming an owner. Nor do men and women who buy municipal bonds come to own the municipality. They acquire no political rights at all. Unless they are already citizens, they cannot even participate in deciding how to spend their own money. They are entitled to a specified rate of interest, and that is their only entitlement. The members of the political community are conceived to have rights of a different sort, simply by virtue of their membership, whether they have invested money in the community or not. There seems no reason not to make the same distinction in economic associations, marking off investors from participants, a just return from authoritative decision making.

Not because men and women join a company voluntarily, with full knowledge of the established structure of authority: the settlers in J-town arrived freely, and the same knowledge was available to them, as J.J. told the citizens' delegation. If I settle in a state founded and ruled by a powerful despot, my knowledge of his despotism does not make the act of settlement into an act of consent. Nor is my prompt departure the only way I can express my opposition to despotic government. That may be the case with a man who joins a monastic order

---

[3] *The Second Treatise of Government*, chap. 5, para. 27.

requiring strict and unquestioning obedience. Here the new member seems to be choosing a way of life, and his choice entails a particular disciplinary rule. We would not be paying him proper respect if we denied the efficacy of his choice; its purpose and its moral effect are precisely to authorize his superior; he can't withdraw that authority without himself withdrawing from the common life it makes possible.

But that is not true of a man who joins a company or who comes to work in a factory. In these associations, the common life does not require unquestioning obedience, and we respect the new member only if we assume that he does not seek subjection. Of course, he encounters supervisors, foremen, company police, as he knew he would, and it may be that the success of the economic enterprise requires his obedience, just as the success of the political association requires that citizens obey public officials. But in neither case do we want to say (what we might say to the novice monk): if you don't like these officials and the orders they give, you can always leave. It is important that there be options short of leaving, connected with the appointment of the officials and the making of the rules they enforce. We see this clearly in the case of towns and also, curiously, in the case of labor unions. In the United States today, the democratic rights of union members are legally protected. But the rights of company employees are only indirectly protected insofar as they are unionized (and insofar as their union has won some share in company decision making).[4]

---

[4] There are other sorts of organizations that raise more difficult problems. What about bureaucracies, for example, or schools? Or consider an example that Marx used (in *Capital*, vol. 3) to illustrate the nature of authority in a Communist factory: cooperative labor requires, he wrote, "one commanding will," and he compared this will to that of an orchestra conductor. A strange comparison, for conductors have historically been tyrannical figures. Should their will be commanding? Perhaps it should, but I doubt that Marx's comparison is a good one, for orchestras must express a single interpretation of the music they play, while patterns of work in a factory are more readily negotiated. The political rights of individuals are relative to the character of the activities in which they voluntarily engage. But this general principle needs to be worked out in detail.

Entrepreneurial vision, capital investment, the freedom to join or not to join: none of these satisfactorily distinguishes economic from political associations. None of them accounts for, let alone justifies, the privacy of a private government.

But there are two differences between towns and companies that I have not yet considered. First of all, a town is an association of residents; a company is an association of workers, who live somewhere else. Perhaps the maxim, *what touches all should be decided by all,* only applies to residential communities. Monasteries and unions are immediate counterexamples, monasteries because they are residential communities to which the maxim does not apply, unions because they are nonresidential communities to which it does apply. Nevertheless, it is true that ordinary democrats have generally tried to organize people where they live, socialists where they work (though not only there), and some distinction might be drawn along these lines. The self-government of residents, it might be said, is more obvious and important than that of workers. Men and women must collectively control the place where they live in order to be safe in their own homes.

There is certainly no other place where it is so important to be safe. Hence another ancient maxim: *a man's home is his castle.* I will assume that this maxim expresses a genuine moral imperative. What does it require? Not self-government, but rather the protection of a private sphere, a piece of nonpolitical space for withdrawal, rest, secrecy, and solitude. As a feudal baron retired to his castle to brood over public slights, so I retire to my home. But the political community is not a collection of brooding places, or not only that. It is a common enterprise, a public place where we are seen and heard by others, where we quarrel over the public interest, where we sometimes work together. That is why the meetings in J-town were so exciting. They represented the discovery or creation of a local republic, a public thing, which by its very nature had to be shared once it was known to exist. And in this sense, an economic enterprise seems to be very like a town, even though, or in part because, it is so unlike a home.

It is not a place of withdrawal, but of cooperative activity. It is not a place that anyone needs to own in order to safeguard his independence and solitude. No one ever thought of saying, a man's factory is his castle. The moral independence of the men and women who work in a factory requires shared decision making and not the protection of a private sphere. Surely we grant that point whenever we require union democracy, and having done that, there seems no principled reason to stop short of company democracy.

But let's think about stopping short, exactly as we currently do. Imagine that the inhabitants of J-town, instead of calling for elections, had organized a citizens' union and bargained collectively with J.J. and his heirs. It is interesting to speculate on the ranges of issues they might have bargained about. Which matters would lie beyond their reach, once they had conceded the issue of ownership? Presumably, they would not have had much to say if J.J. had decided to relocate the town (since it was "his" town). But they could have bargained in detail about living conditions within it, about zoning laws, traffic control, sewage disposal, and so on. They could not have vetoed his choice for chief of police, but it is possible to imagine grievance machinery that would function somewhat like a civilian board of review. The picture is not entirely unattractive, but it is not what we mean by democracy or, at least, it is not all that we mean. Particular groups of city employees do form unions and negotiate with the mayor, but their members also vote for the mayor with whom they negotiate. Members of pressure groups participate in the same dual way, though the arrangements are less formal. They bargain and they vote, acting simultaneously as men and women with particular interests and as men and women with general interests. That seems the right arrangement for economic enterprises also, whose participants are concerned both with their immediate returns and with the well-being of the enterprise as a whole.

The second difference between towns and companies follows from the separation of residence and work. The citizens of a town are also the consumers of the goods and services the town provides—and they are, except for occasional visitors, the only consumers of those goods and services. But the "citizens" of a company are producers of goods and services; they are only sometimes consumers, and they are never the only consumers. Hence there are large numbers of other people, outside the company, who have a direct and material interest in what goes on inside. What should their role be in company decision making? The question is often raised in the literature on workers' control, and it is variously answered. Here I don't want to answer it again, only to insist that the sorts of arrangements required in a fully developed industrial democracy are not all that different from those required in a political democracy. For we don't, after all, grant absolute authority to town governments, even over the goods and services they produce for internal consumption. We enmesh our towns in a federal structure, and we regulate what they can and cannot do in areas like education, criminal justice, environmental use, and so on. No doubt, companies would be similarly enmeshed. In a developed economy, as in a developed polity, different decisions are made by different groups of people at different levels. The division of power in both these cases is only partly a matter of principle; it is also a matter of circumstance and expediency.

The case is similar with the particular constitutional arrangements necessary within companies and factories. There will, of course, be many difficulties in working these out; there will be false starts and failed experiments, as there have been in the previous history of political democracy. Nor should we expect a single resolution of all problems. Proportional representation, single member constituencies, mandated and independent representatives, bicameral and unicameral legislatures, city managers, regulatory commissions, public corporations—the common business is done and should continue to be done in many ways. What is important is that it be known to be common and that our participation in it be recognized as a matter of right.

Today, there are many men and women who preside over enterprises in which hundreds and thousands of their fellow citizens are involved, who make decisions that shape the lives of their fellows, and who defend and justify themselves exactly as J.J. did. I own this place, they say, I built this factory, I founded this company, I risked my capital, I make the decisions around here. It has been my purpose to argue that people who speak this way are wrong. They misunderstand the prerogatives of foundation and investment. They claim an authority to which they have no right. It has not been my purpose, however, to deny the significance of entrepreneurial activity. In both towns and companies one looks for energetic people, willing to innovate and take risks. It would be foolish to create a system that did not bring them forward. They are of no use to us if they just brood in their castles.

On the other hand, nothing they do or can do gives them a right to rule over others—unless they win the agreement of the others. This means that at a certain point in the development of an enterprise, it must pass out of entrepreneurial control. Its founders have created, or they have led other men and women in creating, a public thing, which must now be run in some public way. It is often said that economic entrepreneurs will not come forward if they cannot hope to own the companies they make. The best response is to point to the other side of that all-important but entirely conventional dividing line: we do not lack for political entrepreneurs, though they cannot hope to own the state. Possession is not the goal of public life, but that does not mean that there are not attractive and even compelling goals. For one thing, we can go on building statues of worthy men and women—the founders but not the owners of our common wealth.

(1978)

18

Socialism
and Self-Restraint

The philosophy of *me too* is not hard to understand. Though it has always had its place in social life, especially among the upper classes where competition is most intense, its general force today derives from two related historical tendencies. The first of these is the gradual erosion of social distance and mass deference, and the second is the gradual collapse of lower-class, ethnic, and regional cultures. Social distance insulated and protected the elites; class, ethnic, and regional culture encapsulated the rest of us, focusing our energies inward and limiting the thrust of our ambition. As these two disappear, the world of the elites becomes more knowable and more accessible. From far away, the "better" people really looked better; the princess was always beautiful and (from far away) unattainable. Up close, the illusion fades, and the beauty, *savoir faire*, coolness, and personal force of the few look like mere artifacts: a little luck, a little muscle, a little polish. If her, why not me?

Is this the triumph of envy and resentment? The loss of all

respect for individual merit? Maybe so; but it is also, and more importantly, the triumph of liberalism and capitalism. For these two, reinforcing one another in politics and economy, steadily break down the "Chinese walls" of class and culture. They turn the world, or at least the country, into a single roadway, and they push men and women from every walk of life into a universal competition, the famous "rat race" in which, as Thomas Hobbes wrote in the seventeenth century, there is "no other goal, nor other garland, but being foremost," and where:[1]

> To consider them behind, is glory
> To consider them before, is humility
> To be in breath, hope
> To be weary, despair
> To endeavor to overtake the next, emulation
> To lose ground by little hindrances, pusillanimity
> To fall on the sudden, is disposition to weep
> To see another fall, is disposition to laugh
> Continually to be outgone, is misery
> Continually to outgo the next before, is felicity
> And to forsake the course, is to die.

Caught up in such a race, with eyes fixed on the other runners, it is impossible not to see that merit makes only a part of success; it is impossible to miss the role of luck, the pervasive power plays, the ground that can be gained by a little wit, bombast, deception, meanness. Within the limits of the law—for laws are "as Hedges," wrote Hobbes, "not to stop Travellers . . . but to direct and keep them in such a motion as not to hurt themselves by their own impetuous desires"[2]—anything goes, any subterfuge, any strategem that looks as if it might work. Every man runs for himself.

And devil take the hindmost. Or rather, what is commonly

---

[1] *The Elements of Law*, Part I, chap. 9, para. 21.
[2] *Leviathan*, Part II, chap. 30.

recommended to the hindmost is a little proper respect for the faster runners, loyalty to the system, and resignation to their own fate. It is a matter of deep concern, then, when they assert themselves instead, organize, and demand higher wages or governmental intervention to alter the outcomes of the race. That's what everyone else does, of course; that's what the power plays are all about; and I expect that envy is often the motive. But when the hindmost do it too, the effect is to heat up the race as a whole. Everyone has to run faster simply to hold his place, to maintain what are called (in wage negotiations) "differentials"—the accustomed distance between runners. It turns out that the whole system really does require the resignation of "them behind" and even the carefree and shiftless behavior that follows from resignation. The laggard runners are supposed to fool around, waste time talking to one another, occasionally run amuck beyond the hedges of the law, so that the runners ahead of them can feel superior and secure.

When the race does heat up, prices are inflated, money is devalued, and insecurity spreads from one end of the roadway to the other—though most uncomfortably among the middle runners, where the largest numbers are and the jostling for position is hardest. It feels as if all the discomfort is due to the pressure from behind, and so the cry goes up for restraint, more precisely, for "wage restraint" (or labor discipline). But among "them behind," and especially those who are "in breath," what possible reasons are there for accepting restraint? The others are strung out over such a long distance, the "differentials" are so great, the late starters, the plodders, the slow but steady runners have such a way to go before they rest: why stop now? And if they slow down, what guarantees do they have that the others will slow down too and not increase their own advantage? There are no guarantees, and that's why official calls for wage restraint only produce an intensified scramble, in which those who are able to do so evade or escape the proposed guidelines, and those who cannot sullenly submit—and envy the others: if him, why not me?

The spectacle is unedifying, and the rat race is commonly condemned, both by intellectuals who pretend to be spectators and by acknowledged (and not only "weary") participants. Among free men and women, unconstrained by birth and blood, by class or culture, the race may well be a natural system—though the particular circumstances in which we run and the particular goals we seek (wealth and commodities, say, rather than honor and offices) are historically shaped. Certainly, the running is exciting: it makes for a release of human energies; it makes for large achievements. At the same time, however, it gives people no goal beyond getting ahead, and that is necessarily a goal that cannot be shared with anyone else (except serially: I want to win, and he wants to win, and she wants to win, and so on). In a race, one has competitors, not colleagues or comrades. Hence, "the loneliness of the long-distance runner." And hence, our inability to organize the runners for some cooperative activity.

The race does not, of course, encompass the whole of our lives or provide the only conception that we have of social order. It is a sign of its power, however, that when political leaders want to call us to some common effort, the first alternative imagery they adopt is that of war—as if only crisis and ultimate danger can wrest the runners from their lonely self-absorption. But there are other reasons why the imagery of war makes a plausible alternative to our everyday racing, reasons especially relevant to contemporary politics. Inflation, environmental pollution, energy waste: all these do indeed call for restraint, and restraint may well require, as President Carter has told us, "a moral equivalent of war." But his call to join the fight has only generated the usual scramble—in large part because he has failed to grasp the full meaning of the metaphor he has adopted. He has merely exploited its hortatory power.

War is a useful image for three reasons: first, because it

suggests a common danger; second, because it suggests the need for sacrifice; and third, because it suggests that the sacrifices will be equally, or at least randomly, shared. The last of these is the most important, but let's consider each in turn.

Common danger. Risk taking in the race is an individual business. The defeat of one runner doesn't significantly affect the others. It disposes them to laugh, perhaps, or nervously pump up their own pace; but it's not a common loss. Military defeat is a common loss, and often one that threatens critical as well as widely held values. So self-defense becomes a collective activity, and in its course we experience feelings of solidarity that are alien to the race and the roadway. We are engaged on a new front, where one person's injury or death weakens all the rest.

Sacrifice. On the roadway, sacrifice is not uncommon, but it is narrowly circumscribed by our conceptions of conjugal or parental obligation. "To hold fast by another," wrote Hobbes, "is to love." Sacrifice for a cause or for the sake of one's fellow citizens is hard to imagine, harder to understand. But the solidarity of war creates an emotional climate in which the surrender of self or, at least, of selfishness, is much easier. We can accept austerity without feeling defeated; we can face death without despair. The race produces champions; war produces heroes.

Equality. Danger and austerity will not in fact be shared up and down the roadway, among the rich and the poor, unless the government acts forcefully to reduce the effects of prewar differentials. The most obvious steps are conscription and rationing, both of which aim at distributing the losses of war with the greatest possible fairness. This is "war socialism," and without it, neither solidarity nor sacrifice would be sustained for long. War socialism is almost always incomplete; it is sometimes fraudulent. The race is never entirely broken off, and individual runners still hustle for large and small advantages. After all, they tell themselves, the war won't last forever. But the hustle is, sometimes, subject to new constraints,

more narrowly fixed than the old hedges. And it is often necessary for the government to hold out the hope that war
socialism in one form or another will survive the victory: that
the runners will close ranks in a welfare state, that power will
be shared by winners and losers alike.

It is not my intention to make war seem attractive, for it
obviously doesn't and shouldn't attract us. I only want to suggest what it means to be summoned, in warlike rhetoric, to
solidarity and sacrifice for the common good. My argument
is simple: solidarity and sacrifice won't work without equality.
There can't be a moral equivalent of war unless there is also
a political equivalent of war socialism. One might argue differently in a society where class lines were rigidly drawn and
patterns of deference well established. But not here and now.
We are—all of us—connoisseurs of the roadway, and we
know better than to hold back while others streak ahead.
Short of savage repression, then, there is no way to make
restraint compatible with the usual scramble.

Repression, of course, is always an option: the political
equivalent of civil war. If we decide that the easiest way to
hold down the costs of medical care is to hold down the wages
of hospital workers, then we can defeat the new hospital organizing committees. If urban services seem too expensive, we
can break the municipal unions. If fuel has to be saved, we
can let the costs go up and up, and call in the police to deal
with the protests of the poor and the disadvantaged. If local
movements and governments protest the location of power
plants or the storage of wastes in their vicinity, we can invoke
the joint authority of capital and the state. This is the way the
race was always run, its accumulating costs thrown back onto
"them behind." I suspect, however, that the philosophy of *me
too* is too widely accepted nowadays for this to be as easy as
it once was. The likely victims are too ambitious, too well-
organized, too ready to fight back. Repression is possible, but
it would have to be brutal.

Restraint is better attempted as a cooperative effort. What

does that mean? Not that it has to be voluntary, for no one will cooperate unless he is sure that everyone else is cooperating, and only the law can guarantee that. Sacrifice must be, and must be seen to be, fairly distributed. Hence, first of all, the unavoidability of rationing when necessary goods are in short supply: for there is no other way to distribute them fairly. But rationing is the superimposition of a pattern of equality on an underlying inequality, an equality of (certain) goods on an inequality of income and wealth. The rationing principle does not work when what is at issue is the acquisition of wealth itself. To enforce an "incomes policy" limiting annual increases to, say, 7 percent (when inflation is running at 10 to 12 percent) is to impose very different sacrifices on a person earning $50,000 and on a person earning $8,000. At this point what is called for, if austerity is to be shared, is a reduction of differentials, a policy of redistribution, which might be worked through a strengthening of the welfare system or a rearrangement of the tax system.

This sort of thing is a matter of state action; that's mostly what war is. But it is never successfully fought unless the state acts against a background of patriotic enthusiasm and solidarity. A new spirit of camaraderie must come to shape relations among citizens, reaching its greatest intensity when the common danger is greatest—as among Londoners, for example, during the blitz. We can expect nothing like that, but it is nonetheless important to look for its political equivalent. No federal campaign against inflation or energy waste will have any serious impact unless it generates new patterns of cooperation in towns, factories, hospitals, apartment houses, and neighborhoods. Official plans have to be seconded (but that also means revised and resisted) by groups of citizens helping themselves and one another—restructuring authority relations, redistributing costs, rescheduling work patterns—so as to make common purposes and values more visible than they currently are. All this requires local initiative and popular participation.

The political equivalent of patriotism and wartime solidarity is civism, the citizen's sense of being a participant in a common enterprise. In the writings of contemporary conservatives, civism is frequently discussed, but it looks like little more than the ideology through which self-restraint is to be taught to the working class. And then the term invites suspicion; this is a lesson that won't be learned and shouldn't be. The resulting social order would not have the form of a common enterprise. Civism depends upon equality or, at least, upon much greater equality than we know today. That's why President Carter's call for domestic struggle (like President Ford's WIN campaign) has won so little response. He has not persuaded anyone that the sacrifices required for victory will be equally imposed or that they will result in a more egalitarian society. And if not, why me?

I once thought that the success of the planning state and the welfare state did not require any challenge to the philosophy of *me too*. Planning and welfare were simply the rational instruments of individual desire, and all that was necessary was to make sure that every individual's desires were counted, that there were no pariah peoples, no excluded groups, no permanent under-class. And this men and women mostly did for themselves, in the course of long and often bloody struggles for political recognition and equal treatment by state officials. Solidarity was bred by these struggles and was crucial to their success; but the most general motive was a collective *me-tooism*, and the most common result was individual success, consumer satisfaction, the pursuit of happiness.

New patterns of social cooperation, a deeper equality: these lay, I thought, beyond the welfare state, and waited upon its completion. But this vision assumed an infinitely expandable economy, and that looks today like an implausible assumption. The welfare state can still include everyone— it had better—but its success may well depend upon everyone's restraint. And restraint upon civism, and civism upon equality. The mutual acceptance of limits requires the sharing of what's available within those limits. Without sharing, and without shared decision making, there won't be acceptance, but only evasion, deceit, the old power plays. That means that it is time, now, to think about restructuring the hierarchies of our everyday lives. For the only alternative is the race and the roadway, where scarcity is likely to make things much uglier, the laughing and the weeping much more bitter, than they have been in a long time.

(1979)

ACKNOWLEDGMENTS

Most of these essays were written for the socialist journal *Dissent*, which published my first article in 1956 and from whose editors I received much of my political education. They provided the kind of support and the sense of shared commitment without which I cannot imagine writing about politics at all. In comradely fashion, I blame them for all the incomplete, shortsighted, and wrongheaded positions I have defended in print.

"Civility and Civic Virtue in Contemporary America" first appeared in *Social Research* (vol. 41, no. 1; reprinted by permission of the Graduate Faculty, New School for Social Research, New York, copyright 1979). "Social Origins of Conservative Politics" was published in the now defunct British journal *Views* (no. 6) as an analysis of the Goldwater campaign of 1964. "Nervous Liberals" appeared in the *New York Review of Books* (vol. 26, no. 15; reprinted with permission from *The New York Review of Books*, copyright © 1979 Nyrev, Inc.). "The New Left and the Old" was one of the Frank Gerstein Lectures at York University in Toronto in November of 1967 and was published along with the other lectures for that year in a book entitled *The University and the New Intellectual Environment* (Macmillan of Canada and St. Martin's Press, 1968; reprinted with permission). I have revised it to include some paragraphs from a talk I gave at Harvard the following year and changed the dating accordingly. "The Peace Movement in Retrospect" and "Socialism and Self-Restraint" were written for *The New Republic* (February 10, 1973, and July 7, 1979; reprinted with permission) and represent only a small sampling of the articles and reviews I have published there in recent years. "A Theory of Revolution" appeared in *Marxist*

Acknowledgments

Perspectives (II, 1, 1979); I have added the final section. "Intellectuals to Power?" was published in the *New York Review of Books* (vol. 27, no. 4; reprinted with permission from *The New York Review of Books*, copyright © 1980 Nyrev, Inc.). I am grateful to the editors and publishers of these journals and books for permission to collect the essays here.

All the other pieces appeared in *Dissent* between 1964 and 1979 (copyright © 1964, 1967, 1968, 1971, 1972, 1973, 1974, 1976, 1978, 1979 by Dissent Publishing Association; reprinted with permission). "A Day in the Life of a Socialist Citizen" was also the last of the essays in my book *Obligations* (Harvard University Press, 1970). I am a little uneasy about using it again, but it was written originally for *Dissent*, and it properly belongs to the series of articles on the New Left that makes up Part II of this book. When it first appeared, it carried the subtitle "Two Cheers for Participatory Democracy." It is, anyway, one of my favorites among my own essays.

Except for the changes already noted, I have done only minor stylistic editing, occasionally clarified a clumsy argument, cut out a few verbal repetitions, changed a few titles. The larger repetitions of theme and argument I have not cut; nor have I tried to conceal from the reader the fact that I have sometimes played variations on a theme or made a similar argument in rather different ways, with different emphases. At times I have thought it important to defend liberalism against liberals in rapid retreat from some of its principles; at other times I have criticized those tendencies within liberalism that make for dissociation and weaken the structures of self-government. The introduction, which has not appeared elsewhere, is an attempt at a general statement, but it is also another variation.

Within each section, the essays appear in chronological order, except that I have coupled an early analysis of the social origins of contemporary conservatism with a recent critique of neoconservative intellectuals (chapters 4 and 5).

For the same writers who misunderstood conservatism in the 1960s became its apologists in the 1970s.

I have not included here any of my articles on foreign policy, the Vietnam War, or the Middle East. My early essays on the civil rights movement are omitted; so is most of what I have written over the years about education. The essay on equality in Part 4 is a sketch for a larger argument that I hope to make in a forthcoming book—where some of its points will no doubt be repeated.

INDEX

Index

Corporations, 73–74, 252
Coser, Lewis, 72
Counterculture, the, 63, 150–153, 160–161
Cromwell, Oliver, 218

Debs, Eugene, 65
Decadance, 93
  see also Conservatism, and moral decay
Decentralization, 51–52, 70, 271
  see also Localism
Democratic party, 76, 159, 171–173
Democracy
  and equality, 247, 252
  and public education, 271–272
  and revolution, 222–223
  and socialism, 17, 273–279, 289–290
  in universities, 123, 126
  see also Participatory democracy, Political participation
Deutscher, Isaac, 210n
D'Holbach, Baron, 26
Dissent, 86, 301
Djilas, Milovan, 226, 227
Draft resistance, 121–122

Education
  see Schools, Universities
Encounter, 80, 89
End of ideology, 41–42
Engels, Friedrich, 208
English Revolution, 202, 207, 209
  see also Cromwell, Puritans
Entitlement, 181
Equality
  and bell curves, 238–242
  and conservatism, 101–105, 237–238
  and liberty, 255–256
  and "right reasons," 243–256

and "war socialism," 295–296
  in vanguard ideology, 214–217
  meaning of, 245
Ervin, Sam, 75
Exodus, the, 204–205, 213

Family, 29, 33, 40, 93, 99, 131, 132, 135n, 192
  and school, 260–264
Feudalism, 277–278
Ford, Gerald, 298
French Revolution, 202, 206, 207, 210
  see also Jacobins
Friendship, 12–13, 28, 62, 258

Garrity, Arthur, 267
Ginsberg, Allen, 120
Glazer, Nathan, 94, 101
Goldwater movement, 78–91
Gogol, Nikolai, 31
Gouldner, Alvin, 226–233

Habermas, Jurgen, 52
Hamlet, 73
Harding, Warren, 82
Hedonism, 95–96
Hegel, G. W. F., 191, 229
Hobbes, Thomas, 28, 29, 62, 96, 273, 292, 295
Hofstadter, Richard, 81
Hooker, Richard, 92
Howe, Irving, 86
Humphrey, George, 86
Hunt, H. L., 86
Huntington, Samuel, 94, 96, 100

Income tax, 60
Integration, 163–164
  of schools, 258, 259, 265, 266–269

Index

National Guard, 143
National Mobilization, 170
Nazism, 80
Neighborhood, 47, 99, 176–177, 260–261
Neoconservatism
    see Conservatism
"New class," 102–104, 226–233
New Left
    and community organizing, 113–119, 176–185
    and the counterculture, 150–153, 160–161
    decline of, 157–161
    and "insurgency," 48n
    social origins of, 110–112
    and the university, 122–127, 166–167, 176
    and Vietnam, 119–122, 169–170
    and violence, 139–156
Newark Community Union, 178
Nisbet, Robert, 94, 101, 104
Nixon, Richard, 73, 75, 83, 85, 168, 172
Noblesse de robe, 230

Orwell, George, 152n

Parsons, Talcott, 80
Participatory democracy, 128, 136, 164–165, 178–179
    see also Political participation
Pascal, Blaise, 243–245, 252
Patriotism, 26, 44, 48, 55–57, 67–71, 297
Peace movement, 168–174
Perlman, Selig, 208
Plauralism, 55, 71, 163–164, 258–259, 268, 271–272
Poets, 224–225
Police, 143–146, 148
Polis, 15, 27, 69, 132
Political participation, 28–30, 40, 44

and civic virtue, 64–67
and the New Left, 112–119
in a socialist society, 128–138, 290, 297
    see also Democracy, Participatory democracy
Political parties, 36, 66, 70–71, 76, 115
Pornography, 6, 68, 89
Poujadism, 88
Private government, 277–279, 287
Procedural justice, 16
Professors, 93, 94, 238
Progressive party, 170
Property, 10–11
    private, 215, 277–279, 285, 290
Public Interest, The, 95, 100, 250
Puritans, 203–204, 206–207, 209, 215, 227

Quotas, 253, 255

Racism, 5, 266, 267
Rationing, 297
Rawls, John, 255n
Representation, 35–36, 136, 231–232
Republican party, 78–79, 85
Revolution
    ideology in, 211–217
    Leninist theory of, 217–219
    and modernization, 191–200
    outcome of, 217–219
    vanguardless, 219–223
Robespierre, Maximilian, 56
Rostow, W. W., 189n
Rousseau, Jean Jacques, 54–55, 69, 131–133, 284
Russian Revolution, 202, 209–210, 218
    see also Bolshevism, Lenin, Stalin, Trotsky

Index